Professional Practice in Earth Sciences

Series Editor

James W. LaMoreaux, Tuscaloosa, AL, USA

Books in Springer's Professional Practice in Earth Sciences Series present state-of-the-art guidelines to be applied in multiple disciplines of the earth system sciences. The series portfolio contains practical training guidebooks and supporting material for academic courses, laboratory manuals, work procedures and protocols for environmental sciences and engineering. Items published in the series are directed at researchers, students, and anyone interested in the practical application of science. Books in the series cover the applied components of selected fields in the earth sciences and enable practitioners to better plan, optimize and interpret their results. The series is subdivided into the different fields of applied earth system sciences: Laboratory Manuals and work procedures, Environmental methods and protocols and training guidebooks.

More information about this series at https://link.springer.com/bookseries/11926

Dilber Uzun Ozsahin · Berna Uzun ·
Tamer Sanlidag · James LaMoreaux
Editors

Decision Analysis Applied to the Field of Environmental Health

Springer

Editors
Dilber Uzun Ozsahin
College of Health Medical Diagnostic
Imaging Department
University of Sharjah
Sharjah, United Arab Emirates

Berna Uzun
DESAM Research Institute
Near East University
Nicosia, Turkey

Department of Statistics
Carlos III University of Madrid
Getafe, Spain

Tamer Sanlidag
DESAM Research Institute
Near East University
Nicosia, Turkey

James LaMoreaux
University of Sharjah
Tuscaloosa, AL, USA

ISSN 2364-0073 ISSN 2364-0081 (electronic)
Professional Practice in Earth Sciences
ISBN 978-3-030-96681-2 ISBN 978-3-030-96682-9 (eBook)
https://doi.org/10.1007/978-3-030-96682-9

This Springer imprint is published by the registered company Springer Nature Switzerland AG
The registered company address is: Gewerbestrasse 11, 6330 Cham, Switzerland

Preface

This book uncovers the research areas that cut across multi-disciplines and multi-criteria decision-making analysis in environmental health. The knowledge derived from this book will guide researchers, policymakers, and experts in environmental health to make the most appropriate and accurate decisions and solve problems of uncertainty associated with environmental health. The chapters focus on developing and applying advanced analytical methods geared toward improving effective decision-making policies on health management/treatment processes. Also, it explores other technologies-related advancements based on available information. It improves on them by suitably creating new dynamics of multidisciplinary studies with analysis concluding psychology, statistics, economics, mathematics, public health, management science, architecture, physics, biomedical/medical engineering, and medicine. Finally, the book addresses the development of high-performance medical data collection that would help with effective quantitative modeling and advanced analytical methods to improve decision-making policies by targeting all environmental health-related disciplines and applying multi-criteria decision-making techniques through pure mathematical modeling and artificial intelligence.

Contents

Introduction

Dilber Uzun Ozsahin, Basil Barth Duwa, Mustapha Taiwo Mustapha, and Berna Uzun

Abstract Decision analysis (DA) is a process utilized to evaluate and address the various choices in different research fields. It is a quantitative and qualitative approach to analyzing and addressing the various factors that influence decisions. DA is also applied to tackle environmental health problems in different researches as addressed in this book. Human beings are not independent of the environment; however their environment can affect it. The complexity of the interactions has been identified as a major factor in determining the environmental health of humans. This study focuses on the application of DA in environmental health issues. In this text, we outlined the major factors that are involved using DA applied in the environmental health problems.

Keywords Decision analysis · Decision analysis in healthcare · Decision analysis in environmental health

1 Introduction

Decision analysis (DA) is a formal simultaneous method that selects uncertain variables under certain criteria. In other words, DA characterizes variables based on

---

D. U. Ozsahin
Medical Diagnostic Imaging Department, University of Sharjah, College of Health Science, Sharjah, United Arab Emirates
e-mail: dozsahin@sharjah.ac.ae

D. U. Ozsahin · B. B. Duwa · M. T. Mustapha
Faculty of Engineering, Department of Biomedical Engineering, Near East University, Nicosia, TRNC, Mersin 10, Turkey

D. U. Ozsahin · B. B. Duwa · M. T. Mustapha · B. Uzun (✉)
Center of Operational Research in Healthcare, Near East University, Nicosia, TRNC, Mersin 10, Turkey
e-mail: berna.uzun@neu.edu.tr

B. Uzun
Faculty of Arts and Sciences, Department of Mathematics, Near East University, Nicosia, TRNC, Mersin 10, Turkey

Department of Statistics, Carlos III University of Madrid, 28903 Getafe, Spain

D. Uzun Ozsahin et al. (eds.), *Decision Analysis Applied to the Field of Environmental Health*, Professional Practice in Earth Sciences,
https://doi.org/10.1007/978-3-030-96682-9_1

their discipline efficiencies, costs, qualities, profitability and more. This discipline is applicable in many practical fields, such as in philosophy and in professional to analyze and determine decision necessary in a formal approach.

DA is applicable in many fields such as mathematics, psychology, philosophy, statistics and economy. Therefore, this field could be perceived simply as interdisciplinary. The evolution of the analytical method, where the set of axioms for preference comparisons between decision options occurs and outcomes contains uncertainty which could lead to the expected utility model presented, is traced to 1931 by Frank Ramsey, who pioneered the 'subjective probability' (Fishburn 1989). The first general set of axioms for preference comparisons between acts with uncertain outcomes that leads to an expected utility model was proposed by Ramsey in 1926 and published posthumously.

This made it possible for the birth of much statistical analysis, such as the Leonard Jimme Savage work in 1950 on the axiomatic frame work. Although it is often interdisciplinary, DA has always been considered a branch of studies related to operations research (Keeney and Raiffa 1976). In 1980, the DA society was established to merge operations research with management sciences.

The methodological approach of DA could be applied using framing. This method focuses on the questions and method to characterize variables on decision hierarchy, success, strategy and boundary conditions involved. Similarly, another method in DA that involves representation of uncertain data through probabilities is the quantitative DA (Keeney 2002). The attitude of the decision maker to risk could be reflected through utility functions. Similarly, in terms of conflicting objectives the decision makers attitude to trade-offs could be represented by multi-attribute utility functions (Shachter 1986).

Furthermore, in a prescriptive approach to decision making, an optimal decision based on rational axioms is focused. In a similar approach using descriptive DA research, focuses on explaining people's actual perception on DA, when optimal or preferred. However, some critics showed that formal DA is considered to apathetically affect some researcher's interest in taking decisions instead applying an intuitional feeling to a problem (Fischhoff et al. 1982). Furthermore, for a decision that is made under time pressure, significantly formal methods used in DA have a scanty application with an important expertise and intuition.

Numerous researches have demonstrated the effectiveness of quantitative algorithms in yielding results that have superiority to "unaided intuition". DA has the following basic steps: these are;

i. Definition of decision problem, which is specified by decision-maker/s
ii. Identification of the alternatives
iii. Naming the result of the alternatives
iv. Defining the time horizon
v. Mapping the sequence of events
vi. Determination of the probability of every outcome
vii. Quantifying the values involved

viii. Calculating the expected outcome value of the individual decision of an alternative (Ryder et al. 2013).

Environmental analysis and decision-making are carried out by scientists, policy-makers and society. This works together to make informed decisions that are based on the knowledge and global health outcomes.

Numerous environmental health decisions were made from what were perceived as narrow and disconnected approaches. As a result, many of these decisions were not in a way that maximized the public health benefits (Morton et al. 2009).

A public health approach can be implemented through three key factors: taking responsibility for all possible outcomes, assessing the effectiveness of the interventions, and developing a policy framework that is based on the public. Without a public health approach or systems, trying to solve a problem without first consulting the public can create an unintended consequence (Christopher 2009).

DA is an extremely useful approach that can be applied to a wide variety of challenging issues. It aids in the resolution of problems that are defined by a decision maker between alternatives. It possesses all of the necessary attributes of a beneficial decision assistance tool: It assists us in focusing on what matters, is rational and consistent, and is simple to apply. This book aims to demonstrate the DA's significance in environmental health. This will further broadens the decision-scope maker's for examining numerous relevant issues about environmental health problems in order to arrive at a valuable and beneficial decision.

References

Christopher JP (2009) Environmental Health Sciences Decision Making: Risk Management, Evidence, and Ethics. National Academies Press (US), Washington (DC). Bookshelf ID: NBK50709

Fischhoff B, Phillips D, Lichtenstein S (1982) Calibration of probabilities: the State of the Art to 1980. In: Kahneman D, Tversky A (eds) Judgement Under Uncertainty: Heuristics and Biases. Cambridge University Press, Cambridge

Fishburn P (1989) Foundations of decision analysis: along the way. Manage Sci 35(4):387–405

Keeney R (2002) Value focused thinking: a path to creative decision making. ISBN 0-674-93197-1

Keeney R, Raiffa H (1976) Decisions with Multiple Objectives: Preferences and Value Tradeoffs. Wiley, New York

Morton A, Airoldi M, Phillips LD (2009) Nuclear risk management on stage: a decision analysis perspective on the UK's committee on radioactive waste management. Risk Anal 29:764–779

Ryder HF, McDonough CM, Tosteson AN, Lurie JD (2013) Decision Analysis and cost-effectiveness Analysis. In: Seminars in spine surgery 2007 December vol 21 no 4, pp 216–222. https://doi.org/10.1053/j.semss.2009.08.003

Shachter RD (1986) Evaluating influence diagrams (PDF). Oper Res 34(6):871–882. https://doi.org/10.1287/opre.34.6.87

Theoretical Aspect of Decision Analysis

Dilber Uzun Ozsahin, Mustapha Taiwo Mustapha, and Berna Uzun

Abstract Decision-making has been an important tool for decision-makers, policy-makers, research scientists, government agencies, and parastatals. These have been made easy with the availability of powerful decision-making tools. The tools are easy to grasp and use; hence, they need no extensive training to master. Decision-making tools can carry out various tasks to achieve an appropriate decision-making outcome. It is often used to evaluate decisions made in the context of several variables, and it has a wide range of possible outcomes. With this, realistic and logical outcomes are obtained. Two basic models are used to examine people's decision-making and judgment process; rational decision model and intuitive model. When applied to a real-world problem, decision analysis can become rather complex. The decision-maker must identify the outcomes connected with each decision alternative, assess the probability of each result, and maintain the right sequence of decisions and outcomes. As part of the decision making process, decision analysis uses various methods to assess all relevant information.

D. U. Ozsahin
Medical Diagnostic Imaging Department, University of Sharjah, College of Health Science, Sharjah, United Arab Emirates
e-mail: dozsahin@sharjah.ac.ae

D. U. Ozsahin · M. T. Mustapha
Faculty of Engineering, Department of Biomedical Engineering, Near East University, Nicosia, TRNC, Mersin 10, Turkey

D. U. Ozsahin · M. T. Mustapha · B. Uzun (✉)
Center of Operational Research in Healthcare, Near East University, Nicosia, TRNC, Mersin 10, Turkey
e-mail: berna.uzun@neu.edu.tr

B. Uzun
Faculty of Arts and Sciences, Department of Mathematics, Near East University, Nicosia, TRNC, Mersin 10, Turkey

Department of Statistics, Carlos III University of Madrid, 28903 Getafe, Spain

D. Uzun Ozsahin et al. (eds.), *Decision Analysis Applied to the Field of Environmental Health*, Professional Practice in Earth Sciences,
https://doi.org/10.1007/978-3-030-96682-9_2

1 Introduction

Decision makers, specifically managers and engineers, have understood better that experience, intuition, and judgment are insufficient for effective design and follow-up of necessary processes (Auzefkitap 2021). For this reason, models reflect the role of the decision-maker.

They are very useful tools in designing and carrying out various tasks, as they only bring them to decision-making. In general, two basic models are used when examining people's decision-making and judgment-making processes. One of these two basic models is the rational decision model and the other is the intuitive model. The rational decision-making model is a numerical (quantitative) model while intuitive decision-making model is a verbal (qualitative) model (Auzefkitap 2021). There are many approaches that explain the logical path we follow while making decisions on a chart. Researchers have created many models for rational decision-making. This approach is expressed as an analytical problem-solving method based on a rational decision-making model. The analytical model looks for relationships between inputs or decision variables over which managers have control, and manager's output. It tries to develop numerical definitions that make as much sense as possible for these relationships. The result of these two steps is a model that can be used to evaluate different situations. Also the heuristic used cannot be reduced to a model or a program; but in recent years, we have begun to understand better how we make decisions. Every day we learn new information about intuitive decision making. There are significant differences between these two models in terms of the methods used and the human dimension on which they are based.

The Qualitative model is basically based on intuition, judgment and experience. Behavioral dimensions come to the fore. It is more art than science.

The Qualitative Decision-Making model is often the intersection of cognitive psychology, social psychology, and sociology. This model is based on people. It deals with people's mental and psychological stages while making decisions. It examines the factors affecting these phases and the effects of these factors.

A quantitative model is an approach based on numerical analysis. It includes the establishing models related to the problem that is the subject of the study, based on numerical facts and data. As a result, models come to the fore in analytical thinking or analysis. The quantitative decision-making model is generally the intersection of mathematics, statistics, and engineering approaches. It deals with how to make the most appropriate choice when there are multiple decision alternatives and multiple criteria. It develops models and methods for selection. The environment in which the decision is made is based on the degree of uncertainty in the environment. It minimizes the human factor. It reveals the method required to find a rational solution to decision problems. These methods emphasize how the decision should be made rather than what the decision should be.

Decision analysis (DA) is a methodical, quantitative, and visual way of addressing and analyzing meaningful choices (Decision Analysis (DA) 2021b). It comprises identifying and assessing all of the decision's components before taking steps to ensure the best possible outcome (Decision Analysis (DA) 2021a). Decision analysis guarantees that decisions are made with all pertinent information and options at hand. Numerous businesses and individuals employ this concept while making various decisions, including management, commerce, healthcare, and engineering. As part of the decision-making process, decision analysis uses various methods to assess all relevant information. It is often used to evaluate decisions made in the context of several variables, and it has a wide range of possible outcomes. The advantage of making decisions in a certain setting is that the optimal choice always results in the optimal outcome. Even the ideal decision may not produce the desired result in an unpredictable situation. Thus, there is an uncertainty cost associated with not knowing which event will occur in advance. The processes of decision analysis reflect the process's ultimate objective, which is to choose the optimal alternative. Mainly six steps should be followed in decision analysis as follows (Parnell 2012):

1-Defining the decision situation and objectives.
2-Determining the alternatives.
3-Decomposing and modeling the problem.
4-Obtaining the best alternative.
5-Analyzing the sensitivity.
6-If another analysis is required, return to the first step, if not, the chosen alternative should be applied.

When applied to a real-world problem, decision analysis can become rather complex (Parnell 2012). The decision-maker must identify the outcomes connected with each decision alternative, assess the probability of each result, and maintain the right sequence of decisions and outcomes (Parnell 2012). A decision tree or influence diagram can visualize alternative and feasible solutions and the related risks and uncertainties. A decision tree is a model in the shape of a tree with branches representing possible outcomes. Apart from being intuitive, decision trees provide critical insight into a topic by displaying the outcomes of numerous options and their associated probabilities. As a result, determining which choice results in the best outcome is complicated. A decision-tree analysis is advantageous for organizing these complex decisions. As a result, it serves as a road map for resolving the decision dilemma. Additionally, an influence diagram creates decision models and visual depiction of decision trees.

Whether the decisions are good or not depends on the extent to which they resolve the problems. In addition, whose benefit the decision serves is also an important factor. An effective decision produces the desired outcome as a result of its effect. However, real results will only emerge after decision-executed.

There is no single common way to measure the quality of decisions (Olson and Courtney 1992). For a decision to be effective and efficient, the objectives must be well chosen, the most appropriate means and ways must be found, and it must not contradict the conditions when the decision is finally put into practice. For such qualities to be possible, it is necessary to know all the objectives and means, their qualities, the criteria that will enable the control of these qualities, and the conditions that may affect the decision.

Six decision components determine the quality of the decision; options, knowledge, logic, preferences, environment, and practice. Of these, options, information and preferences form the basis of the decision. On the other hand, information must be reliable and meaningful. The logic built on the foundation should be solid and consistent. For the foundation to be strong, the environment or frame it sits on is also very important. In addition, the decision must be applicable. It is not possible to talk about the goodness of the decision that has no chance to be implemented.

Undesirable consequences may occur at the end of the implementation of the decision. In other words, it is necessary to measure the degree of goodness of the decision not only with the degree of accuracy but also with the degree of total costs incurred, that is, with the degree of rationality. Good decisions can only increase the chances of good results, and in addition, good decisions can lead to wrong results.

The art of good decision-making is formed by systematic thinking (Hammond et al. 1999). The primary key to good decision-making is that the people involved in the decision-making process make decisions using a structured method that combines their preferences and thoughts with relevant information (Kirkwood 1997). A good decision can be defined as using all available resources based on logic, examining all possible options, and making a numerical method. Decision-making activity is future-oriented and based on guesswork. Foresight is the systematic work done to determine the situation that will provide the most benefits, taking into account the long-term expectations of the decision. Every decision takes time, and some experts use that time entirely, some useless. This is at the discretion of the decision-makers. Some problems may occur in the implementation of the decision, and it may be necessary to revise the decision, but if the decision is taken, it is very important that it is applicable and put into practice. The applicability criterion requires being extremely realistic.

Additionally, more sophisticated computer models aid in the decision-making process. With the use of such tools, decision-makers can explore numerous paths to achieving decision-making objectives while simultaneously assessing risks and determining the likelihood that objectives will be realized if outcomes are satisfied. Probabilities are employed to convey uncertainty, whereas trade-offs and utility functions are used to explain conflicting objectives. As a result, objectives are valued in terms of their monetary value (or if the expected values are attained).

Following the development of a model, it is required to compute the expected value (EV) to evaluate which action produces the best outcome. By computing the anticipated value, one can examine the average outcome of all selections and make informed choices. We require the probability of each outcome and the resulting value to estimate the anticipated value. Despite its utility, detractors assert that decision analysis has a basic flaw: analysis paralysis, the inability to decide due to excessive thought. Certain researchers who conduct research on decision-makers' methods also cast doubt on the use of this strategy.

When a real estate development company decides whether or not to build a new shopping mall on a particular site, they may consider a variety of variables. Traffic patterns on various days and hours of the week, the attractiveness of surrounding retail malls, financial demographics, local competitiveness, and purchase preferences are just a few examples. Different simulations can be run using decision-analysis algorithms to assist decision makers in selecting the optimal path among the alternatives.

References

Auzefkitap.istanbul.edu.tr (2021). http://auzefkitap.istanbul.edu.tr/kitap/endustrimuhlt_ue/kararanalizi.pdf. Accessed 22 Oct 2021

Decision Analysis (DA). Corporate Finance Institute (2021a). https://corporatefinanceinstitute.com/resources/knowledge/other/decision-analysis-da/. Accessed 14 Sep 2021

Decision Analysis (DA). Investopedia (2021b). https://www.investopedia.com/terms/d/decision-analysis.asp. Accessed 14 Sep 2021

Hammond K, Keeney R, Raiffa H (1999) Smart Choices: A Practical Guide to Making Better Decisions. Harvard Business School, New York

Kirkwood CW (1997) Strategic Decision Making: Multiobjective Decision. Duxbury Press, Belmont, Analysis with Spreadsheets

Olson DL, Courtney JF (1992) Decision Support Models and Expert Systems. Macmillan, New York, NY

Parnell G (2012) Handbook of Decision Analysis. Wiley

Tools for Decision Analysis. Home.ubalt.edu (2021). http://home.ubalt.edu/ntsbarsh/opre640a/partix.htm#rtreeinflunce. Accessed 15 Sep 2021

Investigating the Effect of Quarry Dust Enhancement on Engineering Behavior of Expansive Soil Using MCDM

Anoosheh Iravanian, Salah Yaseen Al-Dubai, Berna Uzun, and Dilber Uzun Ozsahin

Abstract The volumetric change of expansive soil is considered as a challenge for civil engineers and many methods were produced in order to stabilize such soils. In this study the enhancement of 10, 20, and 30% of Quarry Dust (QD) as stabilization material for four types of silt and clays characterized as expansive soils was explored. The soil samples were collected from different locations in Northern Cyprus with swell percentages ranging from 4 to 20%. The properties of the collected samples were examined in a laboratory and the effect of quarry dust on physical properties, compaction characteristics, one dimensional swell and unconfined compressive strength were measured. Findings displayed that the addition of 10, 20, and 30% QD results in an overall decrease in the Atterberg limits. Addition of quarry dust improved the compaction properties of the obtained soil at all proportions, with the result showing a decrease in the optimum water content and a gradual increase in the maximum dry density. Also, the swell percentages decreased with the increases of the QD while the compressive strength improved. Moreover, the water absorption during the swell decreased with the increase of QD proportion. The results were

A. Iravanian · S. Y. Al-Dubai
Faculty of Civil and Environmental Engineering, Department of Civil Engineering, Near East University, Nicosia, Mersin 10, TRNC, Turkey

B. Uzun (✉)
Faculty of Arts and Sciences, Department of Mathematics, Near East University, Nicosia, Mersin 10, TRNC, Turkey
e-mail: berna.uzun@neu.edu.tr

B. Uzun · D. U. Ozsahin
Center of Operational Research in Healthcare, Near East University, Nicosia, Mersin 10, TRNC, Turkey

D. U. Ozsahin
Medical Diagnostic Imaging Department, University of Sharjah, College of Health Science, Sharjah, United Arab Emirates
e-mail: dozsahin@sharjah.ac.ae

Faculty of Engineering, Department of Biomedical Engineering, Near East University, Nicosia, Mersin 10, TRNC, Turkey

B. Uzun
Department of Statistics, Carlos III University of Madrid, 28903 Getafe, Spain

D. Uzun Ozsahin et al. (eds.), *Decision Analysis Applied to the Field of Environmental Health*, Professional Practice in Earth Sciences,
https://doi.org/10.1007/978-3-030-96682-9_3

then compared by multi-criteria decision analysis (MCDA) based on liquid limit, plasticity index, specific gravity, maximum dry density, optimum water content, clay type, unconfined compressive strength, the sample being naturally occurring and being environmentally friendly by using QD, a waste by-product. The best results were found in the highly plastic silt sample with no enhancement and originally having a low swell potential, followed by the same silt sample with 30% quarry dust enhancement. Among the clay samples the most desired swelling and mechanical behavior was observed in a highly plastic clay soil with original swell percentage of 13.5% in which with 30% QD enhancement it got reduced to 4.4%.

Keywords Quarry dust · Soil stabilization · Expansive clays · MCDA · Compressive strength

1 Introduction

Some clayey soil have seasonal ability for volumetric change due to their capacity to absorb water (Nelson and Miller 1992). However, this volume change induces a ground movement which causes damage and deformation to buildings and roads. Low-rise buildings, and pavements are more exposed to such problems since they don't have adequate weight to resist the exerted pressure. The effect of such phenomena is clearly noticeable in arid and semi-arid zones due to the differences in the amount and the period of precipitation and evaporation (Jefferson 2001). The repeated volume change in expansive soils may cause cracks in buildings and roads, due to swelling-shrinking behavior beneath pavements and foundations. This situation is considered as a challenge in geotechnical engineering and an economical problem for governments due to the extra cost incurred. It was reported that the cost of expansive soil damages in US has equalled the annual average cost of damages by hurricanes, earthquakes, floods, and tornadoes combined (Wyoming Office of Homeland Security 2014).

In order to improve the engineering properties of the expansive soils many methods have been established; Such as mechanical, chemical methods and soil stabilization by additives. In some cases, traditional earth material is more desirable due to its low cost, also industrial by-products can be utilised as superior additives (Basha et al. 2005).

The excessive use of natural resources is becoming a challenge for the environment and society. The substitute materials that once were rejected as waste could be used again to conserve natural resources. One of these by-products is Quarry Dust. The QD is a waste material found at some mining sites accumulated in open areas, and their presences is risky to the surrounding environment causing health problems when inhaled and destroys crops around the mining sites. The current procedure of dumping quarry dust causes environmental problems in contamination of land, groundwater, and the air. This issue is largely due to lack of regulation, lack of monitoring, economic reasons, lack of expertise and risk avoidance. The air emission of the nanoparticles of quarry dust is also very high and it could affect the environment, plants and human health over large areas as it could be spread by the wind over long distances. Exposure to this dust increases during different procedures of crushing, handling and dumping the quarry materials and could cause respiratory diseases for the adjacent people.

Using the quarry dust in soil stabilization of expansive clays of Cyprus could be a suitable solution to stabilize the soil while solving the storing problem of quarry dust in the quarrying sites while minimizing the above mentioned environmental burden and air emissions.

The main purpose of this study is to stabilize four local expansive clay soil samples obtained from North Cyprus-Nicosia with different percentages of QD. The percent addition of QD to each clayey sample were 0, 10, 20 and 30%. Physical and mechanical properties tests are implemented for carrying out the research, which includes Specific gravity test, Hydrometer test, Atterberg limits test (liquid limit and plastic limit), Proctor compaction test, one dimensional swell test under 7 kPa pressure and unconfined compressive strength test. The results of original batches and the stabilized compositions were then compared to decide which of the soil samples would have a better performance and could be a better choice to be used in construction field. The considered properties were; Clay type, Liquid limit, Plasticity Index, specific gravity, maximum dry density, optimum water content, vertical percent expansion and unconfined compressive strength.

2 Materials and Methods

The study was carried out on four expansive soil samples named T1, T2, T3 and T4 which were collected from different parts of North Cyprus to investigate the effect of quarry dust on soils with different expansion ratios. The research was carried out in two stages, the first stage was done on samples without the addition of quarry dust and the second set of tests were carried out with the addition of different proportions of quarry dust, to provide a clear view of the effect of the QD enhancement and also comparison between the results. Quarry dust is a waste material from the quarrying industry as a result of crushing activities and mainly consists of calcium carbonate $CaCO_3$.

There are 14 quarry sites for aggregate production in Northern Cyprus and about half of them are located near residential areas. In these areas the daily production of crushed limestone is about 20,000 tons, but the dust that is produced as a by-product is either just dumped at the quarry site or sometimes used as a filling material in the production of asphalt. Quarry dust is recognized as a detrimental problem, causing environmental harm especially to the areas adjacent the quarry site.

A series of experiments were performed to examine the mechanical and physical properties of soil samples before and after the addition of quarry dust.

The experiments were namely; Specific gravity measurement (ASTM D854), Atterberg limits detection (ASTM D4318), Proctor compression test (ASTM D698), one-dimensional swelling (ASTM D4546), and unconfined compressive strength (ASTM D2166).

Preparation of the specimens was done following the instructions in ASTM code. Soil samples were brought from the field and dried in an oven at 60 °C for 24 h in order to remove the initial moisture content, while avoiding changes in the minerology or Atterberg limits of the samples (Parlak 2011; Huvaj and Uyeturk 2018). Tests such as Atterberg limits, and specific gravity were applied to determine the specimens' physical properties. To determine the mechanical properties, the test samples were prepared at the optimum water content and maximum dry density obtained from the standard Proctor compaction test. All experiments in the research were carried out on both QD treated and untreated samples.

As stated, the aim of this study is to evaluate the original and QD treated soil properties and choose the mixtures that could be more efficient to be used under light weight structures, based on the parameters such as: Specific gravity, clay type, soil expansion prediction by Liquid Limit (LL), degree of expansion regarding the Plasticity Index (PI), soil expansion according to percent swell under 7 kPa, Maximum Dry Density (MDD), Optimum Moisture Content (OMC), Unconfined Compressive Strength (UCS), naturally occurring and using waste by product as shown is Table 1.

For this aim one of the analytical multi-criteria decision analysis (MCDA) techniques, specifically fuzzy preference ranking organization method for enrichment evaluation (f PROMETHEE), was used. Fuzzy logic is an expansion of the Boolean logic and Zadeh (1965) has proposed it in order to define the nonprecise conditions mathematically, furthermore, it supports the decision-makers working with the vague dataset in their analysis significantly (Uzun Ozsahin et al. 2020). It has been enormously beneficial for analyzing the real problems, which are mostly consisting the complex conditions and providing solutions to such problems more rationally compared to classical logic (Uzun Ozsahin et al. 2020).

PROMETHEE is a comparison technique available for comparing and ranking the alternatives, which are containing the conflicting parameters, proposed by Brans et al. in 1984 (Brans et al. 1986). This technique enables decision makers determining the preference functions for each criterion; therefore it can give more sensitive ranking results compared to other MCDA techniques.

Fuzzy PROMETHEE is a combined model for comparing and ranking the alternatives while there are not sufficient or precise inputs of the decision matrix, which is a matrix that contains the parameters of the alternatives Basha et al. (2005). With this method, first, the vague data should be determined by using fuzzy sets or numbers, and then by defuzzification the fuzzy values should be converted to a single value for use in the analysis of the PROMETHEE. In addition to this, using the fuzzy operations the positive and the negative outranking values of the alternatives could be determined as the fuzzy values and at the end of the analysis still, the defuzzification process should be applied for obtaining the net ranking values. There are several techniques of defuzzification including center of gravity, fuzzy mean, the center of area, etc. for different types of fuzzy sets (Brans et al. 1986).

In this analysis, the triangular linguistic fuzzy scale has been used for the determination of the data of the solid types and the importance weights of their criteria.

Table 1 Decision matrix of the soil types

Aim/Importance Weights	Min/VH	Min/M	Min/VH	Min/VH	Max/H	Min/H	Min/L	Max/VH	Max/L	Max/VH
Soil type	Soil expansion according to % swell (Seed et al. 1962)	Clay type (Sridharan and Prakash 2000)	Soil expansion by LL (Chen 1976)	Degree of expansion upon PI (Holtz and Gibbs 1956)	Relative MDD	Relative OMC	Relative specific gravity	UCS (Das 2009)	Naturally occurring	Using waste by product
T1+0%QD	L	M	VH	H	H	VL	H	Stiff	Yes	No
T1+10%QD	L	M	VH	VH	H	VL	VH	Stiff	No	L
T1+20%QD	L	M	H	M	VH	VL	VH	Stiff	No	M
T1+30%QD	L	M	H	H	VH	VL	VH	Stiff	No	H
T2+0%QD	M	H	VH	VH	H	L	VH	Stiff	Yes	No
T2+10%QD	M	H	VH	H	H	VL	VH	Stiff	No	L
T2+20%QD	M	H	VH	H	VH	VL	VH	Very stiff	No	M
T2+30%QD	M	H	VH	VH	VH	VL	VH	Very stiff	No	H
T3+0%QD	M	H	VH	VH	VH	VL	H	Stiff	Yes	No
T3+10%QD	M	H	VH	VH	VH	VL	VH	Stiff	No	L
T3+20%QD	M	H	H	VH	VH	VL	VH	Stiff	No	M
T3+30%QD	L	M	VH	VH	VH	VL	VH	Very stiff	No	H
T4+0%QD	H	H	VH	VH	VL	VH	VL	Medium	Yes	No
T4+10%QD	H	H	VH	VH	VL	M	VL	Stiff	No	L
T4+20%QD	H	H	VH	VH	L	M	L	Stiff	No	M
T4+30%QD	M	H	VH	VH	L	M	M	Stiff	No	H

The fuzzy linguistic scale and their assigned values has been defined as the list as follows:

$$\{VeryHigh(VH)/(0.75, 1, 1), High(H)/(0.5, 0.75, 1), Modarate(M)/(0.25, 0.5, 0.75), Low(L)/(0, 0.25, 0.5), VeryLow(VL)/(0, 0, 0.25)\}$$

Yager index was applied for converting the selected linguistic fuzzy values to defuzzified values. Yager index is an index that can be obtained for the triangular fuzzy values defined by (N, a, b) where N denotes the center, a denotes the distance between the initial bound and the center, b denotes the distance between the terminal bound and the center and it can be calculated using the Eq. 1 American society for testing and materials (2014b).

$$YI = (3N - a + b)/3 \tag{1}$$

After the collection of the dataset of the solid types and selecting the importance levels of the parameters by considering the expert opinion, and applying the Yager index, the PROMETHEE method was applied.

The Gaussian preference functions, that assigns values to alternatives gradually for the ordering of solid types, was selected for each criteria.

In PROMETHEE firstly the preference index of the alternative a_t compared to alternative $a_{t'}$ $(\pi(a_t, a_{t'}))$, must be calculated by the Eq. 2 for each alternative pairs $(a_t, a_{t'} \in A)$, where A denotes the sets of the alternatives

$$\pi(a_t, a_{t'}) = \sum_{k=1}^{K} w_c \cdot [p_c(f_c(a_t) - f_c(a_{t'}))], AXA \rightarrow [0, 1] \tag{2}$$

and w_c denotes the importance weights of the criteria-c and f_c denotes the c-th criteria of the alternatives.

Then the positive and negative outranking flows of each alternative must be determined with the Eq. 3 and Eq. 4 for each criterion.

$$\Phi^+(a_t) = \frac{1}{n-1} \sum_{\substack{t'=1 \\ t' \neq t}}^{n} \pi(a_t, a_{t'}) \tag{3}$$

$$\Phi^-(a_t) = \frac{1}{n-1} \sum_{\substack{t'=1 \\ t' \neq t}}^{n} \pi(a_{t'}, a_t) \tag{4}$$

where $\Phi^+(a_t)$ denotes positive outranking flow and $\Phi^-(a_t)$ denotes the negative outranking flow and n is the number of alternatives.

Then partial pre-order should be determined with PROMETHEE I based on the following statements:

- Alternative a_t is preferred to $a_{t'}$ if;

$$\begin{cases} \Phi^{+}(a_t) > \Phi^{+}(a_{t'})\, and\, \Phi^{-}(a_t) \leq \Phi^{-}(a_{t'}) \\ \Phi^{+}(a_t) = \Phi^{+}(a_{t'})\, and\, \Phi^{-}(a_t) < \Phi^{-}(a_{t'}) \end{cases} \tag{5}$$

- a_t is preferred to $a_{t'}$ if;

$$\Phi^{+}(a_t) = \Phi^{+}(a_{t'})\, and\;\; \Phi^{-}(a_t) = \Phi^{-}(a_{t'}) \tag{6}$$

- a_t cannot be compared with $a_{t'}$ if;

$$\begin{cases} \Phi^{+}(a_t) > \Phi^{+}(a_{t'})\, and\, \Phi^{-}(a_t) > \Phi^{-}(a_{t'}) \\ \Phi^{+}(a_t) < \Phi^{+}(a_{t'})\, and\, \Phi^{-}(a_t) < \Phi^{-}(a_{t'}) \end{cases} \tag{7}$$

net ranking values can be obtained using the PROMETHEE II with the Eq. 8.

$$\Phi^{net}(a_t) = \Phi^{+}(a_t) - \Phi^{-}(a_t) \tag{8}$$

The alternative with a higher positive outranking flow and lower negative outranking flow is the alternative should be preferred when compared to others. Therefore, the alternative with a higher net ranking flow is preferred (Sayan et al. 2020).

3 Result and Discussion

The Specific gravity test was carried out on all soil samples according to ASTM D854 standard. The specific gravity of QD was measured as 2.72. It was observed that the specific gravity of the prepared mixtures increased respectively with the increment of QD addition in all samples as shown in Table 2.

The liquid limit and plastic limit tests were carried out on samples using distilled water and the results are illustrated for all sample in Table 3. It can be seen that before addition of quarry dust the soils were categorized as silt with high plasticity and clay with high plasticity according to the Unified Soil Classification System (USCS). With addition of 10, 20, and 30% quarry dust the liquid limit, plastic limit, and plasticity index decreased over all, but the classification of the soil stayed in the same category. However, there was an incremental increase in the value of LL, PL,

Table 2 Specific gravity for the obtained soil after applying the stabilization method

GS	T1	T2	T3	T4
Soil+0QD	2.55	2.56	2.55	2.40
Soil+10%QD	2.56	2.57	2.57	2.41
Soil+20%QD	2.58	2.59	2.58	2.44
Soil+30%QD	2.60	2.60	2.60	2.47

Table 3 Atterberg limits results for the obtained soils with different percentages of quarry dust

Atterberg limits	Proportions of quarry dust	T1	T2	T3	T4
		MH	**CH**	**CH**	**CH**
LL	0%	64	72	115	132
PL		33	29	32	40
PI		31	44	83	92
		MH	**CH**	**CH**	**CH**
LL	10%	72	62	78	101
PL		34	29	27	35
PI		38	33	51	66
		MH	**CH**	**CH**	**CH**
LL	20%	56	62	59	90
PL		32	27	25	27
PI		24	35	34	63
		MH	**CH**	**CH**	**CH**
LL	30%	57	71	71	97
PL		31	26	26	27
PI		26	45	45	70

and PI at an addition of 10% QD but with more QD added, the values started to decrease.

The standard proctor compaction test method was used to determine the relationship between the water content and the dry density of the compacted soil. It has been observed that the addition of quarry dust improves the properties of clayey soil. With addition of QD and the decrease in the plasticity index PI, the optimum moisture content decreased, while there was an increase in the maximum dry density as shown in Table 4. This response could be due to the higher specific gravity of the QD. The reduction in optimum moisture content of the samples could be related to the replacement of clay minerals with quarry dust grains, which differ from the clay platelets in water absorption capabilities.

Table 4 Results of standard Proctor compaction for the obtained soils

Soil	T1		T2		T3		T4	
	MDD (g/cm^3)	OMC (%)	MDD (g/cm^3)	OMC (%)	MDD (g/cm^3)	OMC (%)	MDD (g/cm^3)	OMC (%)
SOIL+0%QD	1.6	21.5	1.62	22	1.67	19	1.3	39.4
SOIL+10%QD	1.63	21	1.66	21.5	1.74	18	1.37	28.8
SOIL+20%QD	1.7	20.2	1.7	18.3	1.75	18.5	1.4	26.5
SOIL+30%QD	1.72	18.5	1.76	17.5	1.76	18.5	1.44	27.5

The maximum one dimensional swell of each soil sample was measured under a surcharge load of 7 kPa in oedometer cell for having a better understanding of the behavior of the studied samples. The tests were carried out on samples compacted on their maximum dry density and optimum water content, having 50 mm diameter and 15 mm height. The results were noted for each soil till there were no more noticeable increase in swelling measurements. From the results shown in Table 5, it is observed that the swell percentage for all soils was reduced with the addition of QD. The minimum swell value in each soil type was logged for the maximum QD content of that soil sample. However, total one dimensional swell of 10 and 20% QD mixtures for T4 samples illustrate relatively similar outcomes. For the soil T2 with addition of 20 and 30% QD the swell was noticeably reduced, nevertheless the effectiveness of adding these proportions were moderately close having 6.43 and 6.35% swell, respectively. In the other words it could be said that the addition of 20 and 30% of quarry dust to T2 sample results in almost identical trend in their swell behavior.

According to Das (2009), the increase of moisture content reduces the compressive strength in cohesive soil so it could be expected that with reduction of the optimum water content in QD enhanced samples the unconfined compressive strength would increase. Consequently, a noticeable increase was observed in the maximum dry density of all samples with the increase of the content of QD. Similarly an increase in the unconfined compressive strength values for the mixtures was witnessed after the enhancement of the QD, which is a sign of increase of the effective stress in samples. In Table 6, the results of the unconfined pressure test are shown together with the soil consistencies according to Das (2009). As can be seen, the addition of QD improved the consistency of all samples, which could be boldly witnessed in 30% QD contents.

Based on the selected parameters an their importance weight, the results shows that the best three alternative solid types are; T1+0%QD with 0,0030 net ranking value, T1+30%QD with 0,0028 net ranking value, and T1+20%QD with 0,0025 net ranking value (see in Table 7). The last three solid types in the ranking were determined as T4+20%QD with a net ranking value of −0.0033, T4+10%QD with a net ranking value of −0.0049, and T4+0%QD with −0.0052 (see in Table 2). Positive outranking flows and negative outranking flows, which are expressions of the values of strengths and weaknesses of each soil type, were obtained as in Table 7.

Table 5 Swell percent

Soil	Proportion	Swell%	Soil expansion according to % swell (Sridharan and Prakash)
T1	0%	3.80	Low
	10%	2.50	Low
	20%	2.16	Low
	30%	1.69	Low
T2	0%	11.39	Moderate
	10%	10.75	Moderate
	20%	6.43	Moderate
	30%	6.35	Moderate
T3	0%	13.46	Moderate
	10%	9.01	Moderate
	20%	6.68	Moderate
	30%	4.39	Low
T4	0%	19.2	High
	10%	16.34	High
	20%	15.86	High
	30%	12.66	Moderate

Table 6 Consistency and unconfined compressive strength at different quarry dust content

Soil	Proportion	Strength (kN/m^2)	Consistency
T1	0%	133.89	Stiff
	10%	164.02	Stiff
	20%	184.52	Stiff
	30%	197.66	Stiff
T2	0%	194.98	Stiff
	10%	146.44	Stiff
	20%	205.86	Very stiff
	30%	225.94	Very stiff
T3	0%	133.05	Stiff
	10%	151.46	Stiff
	20%	162.34	Stiff
	30%	199.16	Very stiff
T4	0%	66.95	Medium
	10%	128.03	Stiff
	20%	144.77	Stiff
	30%	183.26	Stiff

Table 7 Ranking of the solid types via f-PROMETHEE

Ranking	Solid types	Net ranking values	Positive outranking values	Negative outranking values
1	T1+0%QD	0,0030	0,0049	0,0019
2	T1+30%QD	0,0028	0,0041	0,0013
3	T1+20%QD	0,0025	0,0039	0,0014
4	T3+30%QD	0,0023	0,0038	0,0014
5	T3+0%QD	0,0018	0,0041	0,0023
6	T2+30%QD	0,0017	0,0033	0,0016
7	T2+0%QD	0,0013	0,0038	0,0025
8	T2+20%QD	0,0008	0,0024	0,0016
9	T3+20%QD	0,0005	0,0022	0,0017
10	T1+10%QD	−0,0002	0,0019	0,0021
11	T3+10%QD	−0,0006	0,0016	0,0022
12	T2+10%QD	−0,0008	0,0014	0,0022
13	T4+30%QD	−0,0018	0,0020	0,0039
14	T4+20%QD	−0,0033	0,0013	0,0046
15	T4+10%QD	−0,0049	0,0012	0,0061
16	T4+0%QD	−0,0052	0,0040	0,0091

4 Conclusion

The physical and engineering properties of four local fine grained cohesive soil samples were studied before and after the stabilization with different amounts of quarry dust. Quarry dust which is an industrial by-product was introduced and applied as a stabilizing material into soil to reduce its environmental impact. For four different highly plastic soil samples, the quarry dust addition percentages were decided as 10, 20 and 30% by dry weight. The effect of its appliance was investigated based on liquid limit, plasticity index, specific gravity, maximum dry density, optimum water content, clay type, unconfined compressive strength, the sample being naturally occurring and being environmental by using waste by product and the produced samples were compared by MCDM. Related to the obtained results the conclusion can be listed as follows:

1. The addition of quarry dust resulted in slight increase in specific gravity value of all samples.
2. Atterberg limits (LL and PL) showed and overall reduction with addition of 30% percent QD. This reduction is an advantageous behavior as the Atterberg limits have a direct relationship with the swell potential of plastic soils.
3. The compaction properties of the studied soil samples were improved with the addition of quarry dust. This enhancement was visible in reduction of optimum water content and increase of the maximum dry unit weight.
4. The application of quarry dust to produce a stabilized soil resulted in a significant reduction in swelling behavior of the samples at all rates of application. The total swelling decreased at all ratios, indicating an inverse relationship between the QD substitution rate and the water absorption, where the increase in QD ratio leads to a decrease in the water absorbed during the swelling process.
5. The samples containing higher contents of QD presented higher unconfined compressive strength values.
6. The results of MCDA for comparison between the obtained 16 soil samples showed that based on the physical and mechanical behavior of them, the best properties were obtained in naturally existing T1 soil which is categorized as a highly plastic silt with the lowest one-dimensional swell percent among the untreated samples. In case of using QD by-product the T1 samples with 30 and 20% quantities presented better qualities respectively.

Finally, although there were many benefits observed in addition of quarry dust, it was also detected that using only QD as a stabilizer is not sufficient to decrease the swell capacity efficiently for high swell potential clays. Therefore, it is recommended that other environmentally friendly materials, preferably pozzolanic, would be used together with QD to reduce PI and the swell potential of highly expansive soil.

References

American society for testing and materials (2010) Test methods for liquid limit, plastic limit, and plasticity index of soils. https://doi.org/10.1520/D4318-17E01

American society for testing and materials (2012) Test methods for laboratory compaction characteristics of soil using standard effort (12 400 ft-lbf/ft3 (600 kN-m/m^3)). https://doi.org/10.1520/D0698-12E02

American society for testing and materials (2013) Test method for unconfined compressive strength of cohesive soil. https://doi.org/10.1520/D2166_D2166M-13

American society for testing and materials (2014a) Test methods for one-dimensional swell or collapse of soils. https://doi.org/10.1520/D4546-14E01

American society for testing and materials (2014b) Test methods for specific gravity of soil solids by water pycnometer. https://doi.org/10.1520/D0854-14

Basha EA, Hashim R, Mahmud HB, Muntohar AS (2005) Stabilization of residual soil with rice husk ash and cement. Constr Build Mater 19(6):448–453. https://doi.org/10.1016/j.conbuildmat.2004.08.001

Brans JP, Vincke P, Mareschal B (1986) How to select and how to rank projects. The PROMETHEE method. Eur J Oper Res 24:228–238

Chen Fu (1976) Foundations on expansive soils

Das BM (2009) Principles of geotechnical engineering (Seventh Edition). CL engineering.

Holtz W, Gibbs H (1956) Engineering properties of expansive clays. Trans Am Soc Civ Eng 121:641–663. https://doi.org/10.1061/TACEAT.0007325

Huvaj N, Uyeturk E (2018) Effects of drying on Atterberg limits of pyroclastic soils of Northern Turkey. Appl Clay Sci 162:46–56

Jefferson I (2001) Problematic soils: proceedings of the symposium. Thomas Telford, London

Nelson JD, Miller DJ (1992) Expansive soils: problems and practice in foundation and pavement engineering. Wiley, New York

Parlak M (2011) Effect of heating on some physical, chemical and mineralogical aspects of forest soil. Bartin Orman Fakültesi Dergisi, vol 9

Sayan M, Sarigul Yildirim F, Sanlidag T, Uzun B, Uzun Ozsahin D, Ozsahin I (2020) Capacity evaluation of diagnostic tests For COVID-19 using multicriteria decision-making techniques. Comput Math Methods Med 2020:1–8

Seed H, Woodward R, Lundgren R (1962) Prediction of swelling potential for compacted clays. J Soil Mech Found Div 88:53–87. https://doi.org/10.1061/JSFEAQ.0000431

Sridharan A, Prakash K (2000) Classification procedures for expansive soils. Proc Inst Civ Eng Geotech Eng 143:235–240. https://doi.org/10.1680/geng.2000.143.4.235

Uzun Ozsahin D, Uzun B, Ozsahin I, Taiwo MM, Sani Musa M (2020) Fuzzy logic in medicine. In: Zgallai W (ed) Developments in biomedical engineering and bioelectronics, biomedical signal processing and artificial intelligence in healthcare, pp 153–182. Academic Press

Wyoming Office of Homeland Security (2014) Wyoming multi-hazard mitigation plan. https://trove.nla.gov.au/version/46519567

Zadeh L (1965) Fuzzy sets. Inf Control 8(3):338–353

Superior Types of Bamboo in Healthcare Using with Fuzzy PROMETHEE

Aizhan Syidanova, Huseyin Gokcekus, Berna Uzun, and Dilber Uzun Ozsahin

Abstract The grass bamboo described as the important renewable energy is easy to obtain and represents the value of all forest resources. Bamboo species are in Asia, Europe, America, Africa, and Australia, and Oceania. As the material for paper production, furniture, boats, textiles, musical instruments, and food, and their leaves have also been used as a wrapping material to prevent food spoilage from ancient times. These species accumulate biologically active components in traditional Asian medicine and Ayurvedic medicine to treat various diseases. With the data, the work explains the edibility of bamboo, medical uses, and other uses as household items, furniture, construction, etc. Several tables work together to show what bamboo is

A. Syidanova (✉)
Department of Architecture, Near East University, P.O. Box: 99138 Nicosia, TRNC, Mersin 10, Turkey
e-mail: aizhansyidanova@gmail.com

H. Gokcekus
Department of Civil Engineering, Faculty of Civil and Environmental Engineering, Near East University, P.O. Box: 99138, Nicosia, TRNC, Mersin 10, Turkey
e-mail: huseyin.gokcekus@neu.edu.tr

B. Uzun
Department of Mathematics, Near East University, P.O. Box: 99138, Nicosia, TRNC, Mersin 10, Turkey
e-mail: berna.uzun@neu.edu.tr

D. U. Ozsahin
College of Health Science, Medical Diagnostic Imaging Department, University of Sharjah, Sharjah, United Arab Emirates
e-mail: dozsahin@sharjah.ac.ae

B. Uzun
Center of Operational Research in Healthcare, Near East University, P.O. Box: 99138, Nicosia, TRNC, Mersin 10, Turkey

Department of Statistics, Carlos III University of Madrid, 28903 Getafe, Spain

D. U. Ozsahin
Faculty of Engineering, Department of Biomedical Engineering, Near East University, P.O. Box: 99138, Nicosia, TRNC, Mersin 10, Turkey

DESAM Institute, Near East University, P.O. Box: 99138, Nicosia, TRNC, Mersin 10, Turkey

D. Uzun Ozsahin et al. (eds.), *Decision Analysis Applied to the Field of Environmental Health*, Professional Practice in Earth Sciences,
https://doi.org/10.1007/978-3-030-96682-9_4

capable of when used correctly. In medicine or the food industry, bamboo helps as it contains many beneficial ingredients. For a correct understanding, the MCDM method was used, which is the Fuzzy PROMETHEE was used to organize the tables and display every aspect of the view.

Keywords Bamboo · Types of bamboo · Traditional medicine · Therapeutic potential · Food and beverage · Multi-Criteria Decision-Making (MCDM) · Healthcare use

1 Introduction

Bamboo is an herb well known and used by people in ancient times. Bamboo is the oldest and versatile herb that people apply in variable industries with creative ideas using it. The grass traced back to the pre-pottery epoch, 9,500 years ago (Liese 1999).

Bamboo fibers compare favorably with many others because they have characteristics: hypoallergenic, antibacterial, high wearing resistance as a resistance to fading, hygroscopicity as the capacity to trap ultraviolet rays, antistatic effect, which is natural, has antioxidant potential (Syidanova et al. 2021).

Bamboo grows in the Asian part of the globe, Europe, America, Africa and Australia, as well as Oceania (Syidanova et al. 2021). The herb prefers hot, warm, and humid climates in tropical and subtropical climates. However, many species inhabit a variety of climates which might begin from hot tropical regions to cool mountain and high-mountain forests.

The inter-node regions of bamboo of the stem are generally hollow. The lack of secondary wood causes the stems of monocots, including palms and bamboos, to become columnar rather than tapered.

Bamboo is a flexible and easy-to-use plant and, depending on the treatment, can differ in properties. As a material, bamboo is durable, lightweight and it hardly absorbs water. It is solid and stable, as well as soft and elastic (Syidanova et al. 2021). Bamboo is harvested both from artificially grown in greenhouses and from wild plants. Bamboo species are subdivided, for example Phyllostachys, known as woody bamboo. Bamboo is widely used as a raw material for construction, food, crafts and other industrial products.

1.1 Bamboo as a Plant

There are about 1250 species in about 75 genera of bamboos in the world (Benton 2015). One of the fastest-growing plants on Earth is bamboo. Growth rate depends on local soil and climatic conditions, as well as on grass varieties. Like grass, bamboo stems grow out of the ground to their full diameter and grow to their full height in one growing season lasting three to four months. Each shoot grows vertically, while branches move away from the nodes as they grow. The stem is straight until it reaches the height of an adult. In the following year, the fleshy wall of each stem gradually hardens. In the third year, the stalk hardens even more. About five to eight years to live means the stems are ready to harvest and build in about three to seven years (Krawczuk 2013).

Particular bamboo stalks in subsequent years do not grow larger or taller in diameter than in the first year, and do not substitute growth lost through pruning or natural breaking. Bamboo has a wide range of frost resistance, depending on the type of species and growing area. As the bud and its leptomorph rhizome system, a larger stem forms each year until the plant approaches a certain type of height and diameter limitation (Krawczuk 2013).

1.2 Bamboo Industry Use

In everyday life, due to its ecological properties, bamboo can be found in kitchens in the form of baskets, boards, wooden spoons, plates, which are usually varnished and painted, furniture, keyboards, polished wood speakers, and this is not the limit. Bamboo is used as wood in furniture, sailing ships, carpets, curtains, books, clothing, toothbrushes, even in ornamental plants and many other areas, industrial society uses bamboo in various fields (Syidanova et al. 2021).

About 2 million tons of bamboo shoots uses in cooking each year. In terms of consumption, China is in first place with about 1.3 million tons of bamboo, India is in second place (Ding and Wang 2018). Bamboo is used in Asian cooking and in the preparation of broths and various other dishes, and is available in supermarkets in various forms, both fresh and preserved.

Bamboo rhizomes contain the toxin taxifolin (a cyanogenic glycoside). The secretes cyanide in the intestines, and proper processing makes them edible (Ding and Wang 2018).

Bamboo salt is what Koreans have used for hundreds of years as a traditional medicine method. Korean bamboo salt is known for its therapeutic effect in treating diseases viral diseases, dental diseases, diabetes, circulatory disorders, and inflammatory diseases. A study by Kim et al. (2016) shows that bamboo salt has an immune-boosting effect.

Charcoal is made from pieces of bamboo that have been harvested over the years and burned in ovens at temperatures between 800 and 1200 °C. Bamboo charcoal has a long Chinese history, dating back to 1486 (Huang et al. 2014). It helps protect the environment by reducing residual pollutants.

2 Feature Specifications of Bamboo Parts in Medicine and Food Industries

In medicine and the food industry, bamboo is used partially in the industry where is no need to wait 5–8 years for bamboo to harden, and bamboo can be used even as a root.

Table 1 describes the parts of the bamboo used (Dorr 2004; Nongdam and Tikendra 2014).

3 Material and Methods

The PROMETHEE methodology, developed by Brans et al., aims to comprehensively rank a finite set of alternatives (Uzun et al. 2021). Indicate positive and negative flows that correspond to their evaluation and the level of importance of each criterion (Uzun et al. 2021).

The data in Tables 2, 3, 4 were collected and presented to build a scale from 1 to 5. The collected data enter into the PROMETHEE application for their calculation. Fuzzy baseline data can use to make decisions related to subjectivity, insufficient or partial information, quality information, rough estimation.

Table 2 presents the grade of edibility of bamboo and gives a clear explanation of grass in cooking (Dorr 2004; Nongdam and Tikendra 2014).

Table 3 present the grade of medical uses, rate out of 5, of bamboo. The table gives a clear explanation of using bamboo parts in medicine (Dorr 2004; Nongdam and Tikendra 2014; Wróblewska et al. 2019).

Table 3 present the grade of other uses, rate out of 5, of bamboo and gives clear explanation of parts which can be used and where (Dorr 2004; Nongdam and Tikendra 2014; Wróblewska et al. 2019; Patel 2015; Jamatia 2015).

Table 5 denotes a linguistic scale showing a fuzzy triangular scale with priority for the assessment of criteria. Table 6 presented the bamboo types with their edibility, medicine use, etc. (Useful Tropical Plants, PlantUse, PFAF, Chinese Plant Names Patel).

Table 6: Visual PROMETHEE Application for the Bamboo Types for Medical, Food, and Other Uses shows the numeric scale from previous Tables 2, 3, 4.

4 Results and Discussion

Table 7 Ranking of Bamboo as a Types for Medical, Food and Other Uses with MCDM method with fuzzy logic presented the work of combination of the MCDM method as a PROMETHEE with fuzzy logic to determine which species of bamboo is better. In Table 6, visual PROMETHEE Application for the Bamboo Types for Medical, Food, and Other Uses presented the column of growth, Edibility, Medicine, and Other Uses which is ranging from 1 to 5. Tables 2, 3, 4 show clear explanations of each number. The last column is hardiness which is shows a Celsius degree, the extreme bamboo survival degree. Since bamboo prefers a warm climate, each type of bamboo has its limit in degrees.

Table 1 Feature specifications of bamboo parts in medicine and food industries

Parts of bamboo	Specifications	Collecting	Properties
Shavings	Pale greenish appearance. Sweet	The external side of the bamboo shaves, the middle part of bamboo shaves into long thin slices	Stabilizes body temperature and resolves phlegm; used in acute fevers, helps to deal with vomiting and convulsions
Tabasheer	A semi-transparent white substance. Sweet, cold	Consists of silica and water with traces of lime and potash, acquired from the nodal compounds from certain types of bamboo	Stabilizes body temperature and resolves phlegm; anti-convulsive; used in fever, or loss of consciousness; specifically used in remedies for children's feverish disorders and epilepsy. Prevents coming into being the stones in the gall bladder and kidneys. Bamboo contains a large number of mucopolysaccharides and proteolytic enzymes. Used as a bronchodilator and expectorant. It eliminates inflammatory processes in the respiratory system strengthens and rejuvenates the lung system
Sap (liquid)	The sap is a light yellow color. Sweet, cold	Fresh cut bamboo with superficies removed, as for making shaved stem, is cut but not shaved and heated to discharge the sap from the ends of the parts	Removes heat, resolves phlegm; Helps with febrile illness. Eliminate fever in the lungs and helps with profuse sputum, loss of consciousness
Leaves	It can be bitter, sweet. As a type of long-leaf grass	Grows from bamboo, collected and dried	Clears heat; used in treating fever, urinary retention with blood in the urine, eliminates inflammatory processes
Fresh shoot	Bitter, white	Young shoots are boiled and cut into pieces, shoots must be handled properly, as they contain high levels of toxic cyanogenic glycosides	High in healthy proteins, amino acids, carbohydrates, and many important minerals and vitamins, and very low in fat

Table 2 Grade of edibility of bamboo from 1 to 5

Uses	Rate
Part: Fruit, Leaves **Use:** Fruit. Young leaves-boiled	1
P: Sap, Seed, Shoots, rhizomes, internodes **U:** Young shoots-cooked, boiled, steamed or smoked, and can also be dried or salted for later use. Seeds is cooked and ground into a flour and use for brewing alcoholic beverages. The sap from the stems can be drunk, can be fermented into a wine. From a decoction of the rhizomes makes a refreshing beverage. As a water of good quality can be secured from the internodes	2
Part: Young shoots, Seed, Stem, Sap, Shoots, Fruit, Leaves Uses: Young shoots-boiled like pot-herbs. The seeds are cooked as a substitute for wheat but are small and difficult to harvest in quantity. The sugary sap is made into a drink. Young stems and shoots-cooked as a vegetable. The shoots are slice and dries in the sun for preservation. Large fruits are fleshy and edible. They are used as food from hunger. The leaves used in brewing liquor. Bamboo shoots require boiling in a lot of water or in several changes of water	3
P: Stem, Shoots U: Young shoots-cooked as a vegetable. Large and very palatable when cooked but acrid raw-boil. Not the best quality, but the big size makes them very common. Sleeping young shoots, harvested in winter before they emerge above the ground, are especially tasty as a delicacy. Prepares for consumption by boiling in one change of water	4
P: - U: -	5

The outcome shows that the opening of the first three places in the ranking are occupied by, the most suitable bamboo species: # 1-Phylostacus bambusoides, # 2-Phylostacus nigra punctata, and # 3-Arundinaria gigantea. The result will be changed if the changes are different from the criterion part of the table. The main advantage is the ease of use resulting from linguistic assessments, as well as taking into account the vagueness or ambiguity inherent in a particular topic.

Take a look at Fig. 1, the positive and negative aspects of each technique obtained by fuzzy PROMETHEE. The table in numbers, and the figure, shows visually each type of bamboo and determine its pros and cons.

Consider the first three types of bamboo: # 1-Phylostacus bambusoides by the method for analysis is best in hardiness, edibility, other uses and growth, the medicine is not the best aspect. As for # 2-Phylostacus nigra punctate the hardiness, edibility, medicine use, grow the winning aspect, but other uses are not the best side of this bamboo type. And # 3-Arundinaria gigantea is good for hardiness, other uses edible and growth, but as # 1-Phylostacus bambusoides, the # 3-Arundinaria gigantea is not suitable for medicine. At the top of the picture, the first line is +1, which is a plus for the bamboo subtype, which means bamboo is good at this part. Below you can see −1 this is the part where the bamboo look is not suitable for use. In the middle of the picture is 0, this is the point where all the pros and cons are calculated and the output is displayed which can be seen in Table 7.

Table 3 Grade of medical uses from 1 to 5

Uses	Rate
P: leaves, the siliceous secretion, sheaths of the stem **U:** Cathartic, to stimulate the kidneys, aphrodisiac, and tonic. A decoction of the leaves and nodes and the siliceous matter is used in traditional medicine. Leave antipyretic. New seedlings are used in the treatment of hematuria. Antiemetic, antirheumatic agent. The leaves are used to treat inflammation in arthritis. The stem shells are used to treat nausea and stomach problems	1
P: Shavings of the culm, stems, shoots, bark, leaves, sap, liquid exuded by freshly cut culms, tabashir, roots **U:** Shavings of the stalk cortical substance ('chuk yu') are used in Chinese medicine in the treatment of febrile diseases, hematuria, epistaxis, and infantile epilepsy. The stems help as a remedy for rheumatism. Shoots are used to treat abscesses and malaria. The bark is astringent and helps with menses. The leaves are used to treat heart disease and malaria and are boiled and used in the bath to relieve fever. A decoction of boiled leaves is used by women as a cleanser to expand and scrape off and to facilitate the expulsion of the placenta. The leaves are brewed as hot tea, which induces profuse sweating in case of fever. The sap is used for the treatment of fever and hematuria. The young shoots are used to dissipate the opacity of the cornea. Fluid secreted by freshly cut stems or internodes is considered good eye drops. Young shoots are eaten as a vermifuge. Tabashir, which is a siliceous concretion found in the culms of the bamboo stem, can be collected from the culms. It is used as a tonic in treating respiratory diseases Roots and stems have antitussive, astringent, antipyretic, gastric remedy. The decoction is used in the treatment of fever with flu, acute bronchitis, whooping cough, indigestion, acute gastroenteritis, and tooth abscess, traumatic injuries, pain in bones and muscles. It is especially helpful in treating baby coughs ! There is a risk that overdose could cause respiratory paralysis. Another report says it is toxic, so special care should be taken when using it!	2
P: root, leaves, young sprouts, sap, the epidermis of young stem **U:** Antiemetic, antitussive, astringent, cleansing, diuretic, expectorant, analgesic, sedative, hemostatic. The plant contains large amounts of silica and is in use in Ayurvedic medicine. The root has an astringent and cooling effect. Use to treat joint pain and general weakness. The leaves have an antispasmodic and aphrodisiac effect. They stimulate menstruation and relieve pain during menstruation. They are also for tone up and enhance the function of the stomach; expel worms, and have a reputation as an aphrodisiac. Young shoots are harvests as they grow below soil level and are ingested to relieve nausea and indigestion. As a poultice, it helps drain infected wounds. The juice is rich in silica, taken of osteoarthritis and osteoporosis internally to strengthen the cartilage tissue. The leaves have antipyretic and diuretic effects. They are for the treatment of fevers, especially childhood seizures, epistaxis, and vomiting. The juice of the stalks lowers fever, has an antitussive, anti-inflammatory, and soothing effect. It is taken orally for the treatment of lung infections with cough and phlegm. The epidermis of the trunk bark has an antiemetic, cleansing, and soothing effect	3
P: - U: -	4
P: - U: -	5

! Please consult your doctor before using the plant for medicinal purposes. This work is not responsible if you decide to be treated by the folk methods!

Table 4 Grade of other uses from 1 to 5

Uses	Rate
P: canes, internodes, rhizomes, reeds, internodes **U:** Containers, pipes, timber, hedges, reeds, used as household utensils, etc. Short internodes at the bottom of the reed are for flasks, vase, etc. Heavy construction, weaving, needlework, and paper making. Rhizomes are used to make handles for umbrellas and canes. Wicker products and musical instruments are also made	1
P: Culms, split canes, **U:** The culms is used for agricultural equipment, as well as pillars and scaffolding. Split canes used in weaving and handicrafts. Culms is used for the manufacture of rough baskets and light furniture like a table, chair, etc.	2
P: Stems, culms, dried canes, root **U:** The plant is suitable for the reclamation of ravine land. The stems are used for-scaffolding, bridges, poles, agricultural implements. They are also split and used to walking sticks, furniture and baskets. The stems are used as a raw material in paper mills. Long, straight stems in rough structures as a frame; in the fishing industry, made into rafts, fish traps, outriggers, and fish pens; as temporary water conduits; in a fence, etc. Culms are for furniture walls, divided for weaving walls, used in handicraft production The extensive root system makes it suitable for erosion control in land rehabilitation. The strong woody culms used as a building material for huts, for making fencing, musical instruments, xylophones, and tambourines, etc. The dried canes are used as a fuel and are also made into charcoal	3
P: root, stems, internodes **U:** the root system is good for helping to control soil erosion. Extensive plant growth provides stabilization of the river bank, delayed sedimentation and bioaccumulation of nutrients and toxins, the plant makes an excellent, dense hedge or screen. Canes are used as tubular stems, woven into baskets and rugs, plus many other purposes. Hollow stems can be turned into grooves. Carbon Agriculture Solutions-Industrial Crops: Crops grown for non-food purposes. Traditional materials include lumber and straw, paper and cardboard, and textiles. The pungent smoke generated by burning the stem is used as a mosquito repellent, the stems serve as supports for banana plants, used for light construction work, crafts, irrigation pipes, trellises, bridges, dwellings, furniture, boat masts, etc.	4
P: Culms, stems, culm, internodes, leaves **U:** For windy sites, it's planted to make windbreaks around farms. Plants along rivers to see floods. The stems are produced as scaffolding, rafts, furniture, and paper. Nearly all furnishings in the home, including mats, screens, chairs, tables, bed frames, and bedding, are frequently made of stems, household utensils. All parts are used in shipbuilding, the construction of bridges, water pipes, ship sails, masts, and rigging. Buckets, jugs, jars, and cups are made up of the stem section. Baskets, boxes, fans, hats, and jackets are made from split bamboo stems. Ropes and Chinese paper are made up of fibers inside the stems. Agricultural implements of all kinds, for spinning cotton and wool or winding silk, are often made entirely of bamboo. The leaves for packaging, as a filler for mattresses, etc.	5

There are two # 11 in table 7, Melocana baccifera and Gigantochloa levis, data is similar, hardiness is not the best part for both types of bamboo, as shown in Fig. 1, but from Table 6 both species have −1 hardiness and there are several other types with −1. The tables are related to each other, so each table is important to understand each type of bamboo.

Table 5 Linguistic scale for importance for the bamboo types for medical, food and other uses

Linguistic scale for assessment	Triangular fuzzy scale	Significance of criteria ratings
Very high (VH)	(0.75, 1)	Medicine, Edible
Important (H)	(0.50, 0.75, 1)	Other uses
Medium (M)	(0.25, 0.50, 0.75)	Hardiness
Low (L)	(0, 0.25, 0.50)	Growth

Table 6 Visual PROMETHEE application for the bamboo types for medical, food and other uses

Scientific name	Growth	Edible	Medicine	Other uses	Hardiness
Arundinaria gigantean	F	3	1	4	−10
Bambusa bambos	F	3	3	5	−1
Bambusa tuldoides	F	2	2	2	−7
Bambusa vulgaris	F	3	2	4	−2
Dendrocalamus giganteus	F	2	1	4	−1
Dendrocalamus strictus	F	3	1	3	−5
Dinochloa dielsiana	F	2	2	2	−1
Dinochloa scandens	F	2	2	2	−2
Gigantochloa levis	F	3	2	3	−1
Melocana baccifera	F	3	2	3	−1
Nandina domestica	M	1	2	1	−9
Oxytenanthera abyssynica	F	2	2	3	10
Phylostacus bambusoides	F	4	1	4	−15
Phylostacus edulis	F	4	1	1	−4
Phylostacus nigra	F	4	3	1	−5
Phylostacus nigra henonis	M	4	3	1	−9
Phylostacus nigra punctata	F	4	3	1	−18
Schizostachyum glaucifolium	F	0	2	3	−2

Growth: F—Fast, M—Medium
Edible, Medicine, Other Uses—out of 5 (In comparison with other plants that can and cannot be eaten, because of that there are no numbers 4 in Edibility, 4 and 5 in Medical use, the same applies to the edibility of medicine table)
Hardiness—Celsius Degree, extreme bamboo survival degree

For the last three, # 15 is Dinochloa dielsiana, # 16 is Schizostachyum glaucifolium, # 17 is Oxytenanthera abyssynica, all three medicinally suitable and they are fast-growing species, # 16 is Schizostachyum glaucifolium, and # 17 is Oxytenanthera abyssynica their plus is other uses, other use is limited in all aspects, so they are in the last place.

Table 7 Ranking of bamboo as a types for medical, food and other uses with MCDM method with fuzzy logic

Rank	Alternatives	Phi	Phi+	Phi−
1	Phylostacus bambusoides	0,1776	0,2019	0,0243
2	Phylostacus nigra punctata	0,1614	0,2053	0,0439
3	Arundinaria gigantean	0,1054	0,1458	0,0404
4	Phylostacus nigra henonis	0,0954	0,1563	0,0609
5	Bambusa tuldoides	0,0312	0,0858	0,0545
6	Phylostacus nigra	0,0268	0,1070	0,0802
7	Bambusa bambos	0,0215	0,0922	0,0707
8	Dendrocalamus strictus	0,0112	0,0733	0,0621
9	Bambusa vulgaris	−0,0077	0,0581	0,0658
10	Nandina domestica	−0,0198	0,1002	0,1200
11	Gigantochloa levis	−0,0339	0,0409	0,0748
12	Melocana baccifera	−0,0339	0,0409	0,0748
13	Phylostacus edulis	−0,0388	0,0687	0,1075
14	Dinochloa scandens	−0,0649	0,0233	0,0881
15	Dendrocalamus giganteus	−0,0671	0,0415	0,1085
16	Dinochloa dielsiana	−0,0756	0,0209	0,0965
17	Schizostachyum glaucifolium	−0,1357	0,0293	0,1650
18	Oxytenanthera abyssynica	−0,1530	0,0222	0,1752

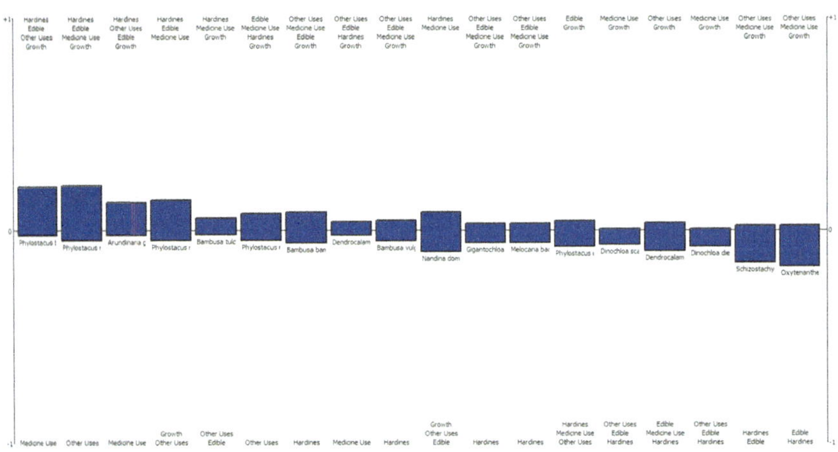

Fig. 1 Positive and negative aspects of each technique obtained by fuzzy PROMETHEE

5 Conclusion

Bamboo is the grass that people have been using in their home life for many years now. Bamboo has around 1250 species, day by day industries getting related to bamboo. And, not all of them are usable in medicine, the food industry, and other uses as the crafting of house utensils or as furniture. The fuzzy PROMETHEE method was used for data analysis, Phylostacus bambusoides, Phylostacus nigra punctata, and Arundinaria gigantea is the top in the ranking. Research shows that Phylostacus bambusoides and Arundinaria gigantea are not that good for the medical industry, only leaves, the siliceous secretion, sheaths of the stem can be used from this species. However, these species are advantageous in the food industry, stems, culm, internodes, leaves are well for scaffolding, rafts, furniture, paper, household utensils, all kinds of agricultural implements. For Phylostacus nigra punctata is not good in other uses and can be used only for heavy construction, pipes, and household utensils as spoons or as containers, the rhizomes are used to make handles for umbrellas and canes. On the other side, the Phylostacus nigra punctata is useful in medicine and efficient in the food industry.

The research showed that 2 billion bamboos in the food industries. Through the centuries, bamboo comes as a product that is growing fast and ecologically perfect. Some types have antioxidants which are superior in medicine, traditional medicine. In Asian countries bamboo is used as a traditional treatment in medicine.

For a proper understanding of the bamboo, the Fuzzy PROMETHEE method was used to organize each aspect of the species. The tables with data can give a clear view of few species of bamboo, in which parts their propriety is good, even examples of where exactly they can use. When used correctly in medicine or cosmetics, bamboo helps as it has many beneficial ingredients.

References

Bellman RE, Zadeh LA (1970) Decision-making in a fuzzy environment. Manage Sci 17(4):141–164. https://doi.org/10.1287/mnsc.17.4.b141

Benton A (2015) Priority species of bamboo. In: Liese W, Köhl M (eds) Bamboo. TROPICAL, vol 10. Springer, Cham. https://doi.org/10.1007/978-3-319-14133-6_2

Chinese plant names, 26 March 2020. http://www.efloras.org/flora_page.aspx?flora_id=3

Ding M, Wang K (2018) Determination of cyanide in bamboo shoots by microdiffusion combined with ion chromatography–pulsed AMPEROMETRIC detection. R Soc Open Sci 5(4):172128. https://doi.org/10.1098/rsos.172128

Dorr SDC (2004) Active constituents. Bamboo as medicine, December 2004. http://www.itmonline.org/arts/bamboo.html, 11 Sept 2021

Huang P-H, Jhan J-W, Cheng Y-M, Cheng H-H (2014) Effects of carbonization parameters of moso-bamboo-based porous charcoal on capturing carbon dioxide. Sci World J 2014:1–8. https://doi.org/10.1155/2014/937867

(2007) Introduction to multiple attribute decision-making (MADM) methods. In: Decision making in the manufacturing environment. SSAM. Springer, London. https://doi.org/10.1007/978-1-84628-819-7_3

Jamatia S (2015) Livelihood of the bamboo base: challenges and opportunities. http://www.academia.edu/3794654/Livelihood_of_the_Bamboo_base_Challenges_and_Opportunities
Kim N-R, Nam S-Y, Ryu K-J, Kim H-M, Jeong H-J (2016) Effects of bamboo salt and its component, hydrogen sulfide, on enhancing immunity. Mol Med Rep 14(2):1673–1680. https://doi.org/10.3892/mmr.2016.5407
Krawczuk K (2013) Bamboo as sustainable material for future building industry, KEA-Københavns Erhvervsakademi, October 2013
Liese W (1999) Bamboo: past–present–future. Am Bamboo Soc Newsl 20(1):1–7
Nongdam P, Tikendra L (2014) The nutritional facts of bamboo shoots and their usage as important traditional foods of Northeast India. Int Sch Res Not 2014:1–17. https://doi.org/10.1155/2014/679073
Ozsahin DU et al (2019) Evaluation and simulation of colon cancer treatment techniques with fuzzy PROMETHEE. In: Advances in science and engineering technology international conferences (ASET). https://doi.org/10.1109/icaset.2019.8714509
Ozsahin I, Sharif T, Ozsahin DU, Uzun B (2019) Evaluation of solid-state detectors in medical imaging with fuzzy PROMETHEE. J Instrum 14(01):01019. https://doi.org/10.1088/1748-0221/14/01/c01019
Patel A (2015) Bamboo structures. School of Mechanical, Aerospace and Civil Engineering
PFAF, 26 March 2020. https://pfaf.org/user/Default.aspx
PlantUse, 26 September 2020. https://uses.plantnet-project.org/en/Main_Page
Schröder S (2012) When and how to harvest bamboo, 15 November 2012. https://www.guaduabamboo.com/cultivation/when-and-how-to-harvest-bamboo, 18 Apr 2018
Syidanova A, Gokcekus H, Uzun Ozsahin D (2021) Superior types of bamboo as a construction material with MCDM methods. In: Uzun Ozsahin D, Gökçekuş H, Uzun B, LaMoreaux J (eds) Application of Multi-Criteria Decision Analysis in Environmental and Civil Engineering. PRES. Springer, Cham. https://doi.org/10.1007/978-3-030-64765-0_10
Useful Tropical Plants, 26 March 2020. http://tropical.theferns.info/
Uzun B, Almasri A, Uzun Ozsahin D (2021) Preference ranking organization method for enrichment evaluation (PROMETHEE). In: Uzun Ozsahin D, Gökçekuş H, Uzun B, LaMoreaux J (eds) Application of multi-criteria decision analysis in environmental and civil engineering. PRES. Springer, Cham. https://doi.org/10.1007/978-3-030-64765-0_6
Wróblewska KB, de Oliveira DCS, Grombone-Guaratini MT, Moreno PRH (2019) Medicinal properties of bamboos. Pharmacognosy-Medicinal Plants. https://doi.org/10.5772/intechopen.82005
Zadeh LA (1965) Fuzzy sets. Inf Control 8:338–353

Evaluation of the Green Campus and Sustainable Campus: Green Building Rating System and Sustainability Approach in Higher Education

Aizhan Syidanova, Çiğdem Çağnan, and Dilber Uzun Ozsahin

Abstract Campuses playing a big role in the eco-system of the city since the university it's the main place for students, scientists, teachers, and academicians. It's important to show how to help the environment from the beginning and make it a basic rule for the future, something common and must have in our lives.

The green building rating system such as LEED, BREEAM, WEEL, Fitwel, Green Globes, DGNB, HQE, Green Star, and Sustainable Assessment Rating Systems in Higher Education such as SAQ, Penn State Indicators Report Approach, STARS, ASSC, Greening Universities Toolkit was examined and based on literature review obtained from books, websites, and articles. Explanation and discussion of the difference between sustainable and green building for proper understanding of the concept of what university campuses needs.

Keywords Green campus · Sustainable campus · Green building rating system · Sustainability approach in higher education

1 Introduction

Modern university campuses are a mini-city like a city should be in the ecosystem of the place itself, without disturbing the balance. Campuses are not part of nature, but without polluting the environment or even helping nature, the campus can be

A. Syidanova (✉) · Ç. Çağnan
Department of Architecture, Near East University, Nicosia, TRNC, Turkey
e-mail: aizhansyidanova@gmail.com

Ç. Çağnan
e-mail: cigdem.cagnan@neu.edu.tr

D. U. Ozsahin
Medical Diagnostic Imaging Department, University of Sharjah, College of Health Science, Sharjah, United Arab Emirates
e-mail: dozsahin@sharjah.ac.ae

Center of Operational Research in Healthcare, Near East University, P.O. Box: 99138, Nicosia, TRNC, Mersin 10, Turkey

D. Uzun Ozsahin et al. (eds.), *Decision Analysis Applied to the Field of Environmental Health*, Professional Practice in Earth Sciences,
https://doi.org/10.1007/978-3-030-96682-9_5

viewed as an autonomous model. The definition of a green campus is not simple as it sounds, a place for people in higher education, the system of knowledge plus the environmentally friendly practices and how they influence each other. The practice of sustainability, look for the future of the campus, developments, and technologies in this area in the design and construction of university campuses also require the method or step-by-step operation plan.

Universities are the best places for researching sustainability and green buildings. Higher education has become a part of teaching and providing information about these two fields. Higher education began to be interested in the environment as early as the 1970s, and academicians began to realize that the environment was threatened by social and economic impacts (Finlay and Massey 2012). The green campus strategies are required to connect green curricula, greenish policymaker, energy conservation, air, recycling, transportation, planning and design, greenish labs, greenish information technology, and learning and teaching (Fachrudin 2020). Nine criteria for a sustainable campus are energy, food, materials, governance, investment, wellness, curriculum, interpretation, and aesthetics (Thomashow 2014). The target of investing in greening places is, recycling will save on disposal costs. Investing in energy will save on electricity costs, investing in saving and recycling the water. Using green design as a green improvement method will improve the living standards of students and campus staff.

The principles of green campuses there are the seven main points: building, water conservation, energy conservation, landscape, waste management, transportation, education (Fachrudin 2020). The priority of land using in ecocity which can be used in universities, revising transportation problems, restore the urban environment, housing problems, support green projects, create gardening, promote recycling, reducing all types of waste, resource conservation, the priority of educational projects for the local environment and provide economic and ecological activity towards pollution, waste and material goods (Finlay and Massey 2012).

Sustainable and Green standards in higher education mostly talking about the things close to each other, however, not all green can be sustainable, and not all sustainable can be green. That's why Green Building System exists, to create a guideline with a tool that will show where exactly people may use it or how.

1.1 Green Building Rating System

The green concept of university campuses is focusing on innovation, discussing environmental issues, researching green buildings in universities, creating space that can contribute to a sustainable environment.

More than a hundred building certifications rating tools exist in the world. The green building systems are guidelines and certifications for improving sustainability in the building industry. The system explains the way of construction, designing buildings, operation, of reducing the negative impact on the environment.

Table 1 Overall information of green building rating system, the ranking system which working in few countries and they are expanding their boundaries

Rating system	National	Countries	Establishing year	Total amount of certified building	Official website
LEED*	USA	160	1998	79,000+	http://leed.usgbc.org/leed.html
BREAM*	UK	93	1990	598,201	https://www.breeam.com/
WELL*	USA	60+	2014	4000	https://www.wellcertified.com/
Fitwel	USA	40+	2012	2400	https://www.fitwel.org/
Green Globes	USA/Canada	83	1997		https://greenglobe.com/about/
DGNB*	Germany	29		5000+	https://www.dgnb.de/en/index.php
Cerway (HQE*)	French	25+	2013	380 000+	https://www.behqe.com/cerway
Green Star	Australia		2003	1,300+	https://new.gbca.org.au/

*Noted:
LEED - Leadership in Energy and Environmental Design
BREAM - Building Research Establishment Environmental Assessment Method
WELL - The Well Building Standard
DGNB - Deutsche Gesellschaft für Nachhaltiges Bauen
HQE - Haute Qualité Environnementale

Each green building rating system has its number of main principles. Each one of the principles is designed for its own country and climate zone. However, with the growing world, some of the rating systems expanded to the international market.

World Green Building Council (WGBC) was established in 1993, a global network that is leading sustainable changes in the building industry to make our environment healthier.

The ranking tools recognize and reward companies and organizations that build and operate greener buildings. Rating tools using the market to set standards, which in turn raise the ambitions of government building codes, staff training, and corporate strategies.

Assessment tools differ in their approach. And they can also vary by building type, with specific approaches or subsets of tools used for different buildings, commercial buildings, or even entire neighborhoods (Table 1).

1.2 Sustainable Assessment Rating System in Higher Education

The United Nations World Commission on Environment and Development prepared a concept called Brundtland Report in 1987 as a Sustainable Development (Federal Office for Spatial Development ARE 2021). Nowadays, the word "sustainability" using in a wide way and can be included in different aspects of human life. Numerous declarations promoting sustainability in educational institutions from the 1990s to the 2000s; strengthened the university management initiatives to integrate sustainable development issues, especially in Europe and America (Federal Office for Spatial Development ARE 2021).

These countries have established councils of higher education institutions; in particular, in 1996 was founded the Environmental Association of Universities and Colleges (EAUC) of the United Kingdom and Ireland, in 2006 were founded the global network International Sustainable Campus Network (ISCN) and Association for the Advancement of Sustainability in Higher Education (AASHE) of the United States of America and Canada (Hokkaido University Sustainable Campus Management Office n.d.). The pioneering universities in these countries can be said to have pioneered the Sustainable Campus program.

Nowadays, promoting the concept of Sustainable Campus has become a responsibility of different higher education institutions in the United States, Europe, Canada, and other countries. For example, in 2010, Hokkaido University in Japan established its Sustainable Campus Office and supported campus sustainability initiatives among Japanese universities. A new system was reorganized into the "Office for Sustainable Campus Management" on April 1, 2018.

June 23, 2021, the federal Office for Spatial Development ARE declared the 2030 sustainable development strategy; all United Nations member states made a political commitment to achieving the 17 Sustainable Development Goals (Federal Office for Spatial Development ARE 2021). The obligation was national and international and accepting the responsibilities for the present and future generations. The Federal Council stated that current problems such as a pandemic, climate change disasters, economic crises, and conflicts are the general awareness and needs in changes towards sustainable development (Federal Office for Spatial Development ARE 2021).

A sustainable university usually aims to the whole institution, targeting to simultaneously influence it through various leverage, such as (Beringer and Adomßent 2008):

- In fields of activity: teaching, research, advocacy or knowledge transfer, community service,
- Academy and Administration/operations;
- The system of public policy of higher education (macro-level), institutional management and administration (Meso level) and operating (micro-level);
- Different points of view of the technical, systemic, and paradigmatic.

Table 2 Sustainable assessment rating system in higher education—sustainability standards in higher education

Assessment tool	Organization	Country	Date of development	Overall information	Official website
SAQ*	ULSF*	UK	1999–2001	The Quantitative Questionnaire provides people-to-people dialogue about university sustainability, a comprehensive definition of sustainability	http://ulsf.org/sustainability-assessment-questionnaire/
Penn State Indicators Report Approach	Penn State Green Destiny Council	USA	2000	Quantitative Questionnaire main points are Water, Energy, Material, Transportation, Built Environment, Community, Research, and Decision-making Food, Land	http://equity.psu.edu/indicators/list-indicators https://p2infohouse.org/ref/17/16964.pdf
STARS*	AASHE*	USA	2007–2010	Understanding sustainability in all parts of higher education Organize improving activities towards sustainability Help to exchange information on how to show sustainability and performance in high education. Forming a strong and diverse community for sustainable campus development	https://stars.aashe.org/

(continued)

Table 2 (continued)

Assessment tool	Organization	Country	Date of development	Overall information	Official website
ASSC*	Hokkaido University	Japan	2013	Questionnaires results are examined closely regarding selected evaluation standards that are needed to realize the sustainability of a campus Four categories of assessment: administrative, environmental, education / research, and regional society, and its mechanism is such that from the assessment result, policies of administration can be seen, in which area reinforce its strength or to reduce weakness	https://www.osc.hokudai.ac.jp/en/action/assc
Greening Universities Toolkit	UNEP's EETU, UN, experts, researchers, GUPES**	Australia	2011	Provide universities with strategies and tactics to tackle climate change, improve resource efficiency, improve ecosystem management, and minimize waste and pollution. The focus is on sustainable planning, design, development, and campus management	https://www.unep.org/resources/toolkits-manuals-and-guides/greening-universities-toolkit-v20

* SAQ - Sustainability Assessment Questionnaire
*STARS - Sustainability Tracking and Assessment Rating System
*ASSC - Assessment System for Sustainable Campus
*ASSC - Assessment System for Sustainable Campus
*ULSF - University Leaders for a Sustainable Future
*AASHE - Association for the Advancement of Sustainability in Higher Education
**United Nations Environment Programme (UNEP)'s Environmental Education and Training Unit (EETU) in partnership with other United Nations agencies and leading "green universities" experts and researchers, under the umbrella of the Global Universities Partnership for Environment and Sustainability (GUPES)

It aims to systematically, strategically, and holistically promote sustainable development on the campus. For collective learning about sustainable campuses, scientific research helps in guiding and supporting in reaching the target.

Table 2 provide overall look on some of the sustainable assessment tool for higher education, in the part of overall information there is what tools is aiming for and what they providing.

Most of the tools were to increase knowledge of sustainability in higher education and create awareness of how to prevent waste management, giving a look to reducing greenhouse gas emissions, etc. Some of the tools created for people to ask more questions, be more goal-centered toward their institution and location of the particular institution, and make them create decisions towards better-using campuses with a green and sustainable approach.

All the rating tools divided by subsystems, and with their subtypes as administration or management, and academy work (research, learning, knowledge transfer, community work, and services) which interact with each other in difficult and confusing ways that are understood more than anything else in the systems (Beringer and Adomßent 2008). Like every other system is more than its constituent parts; synergy—for example, between administration and operation process; study and teaching; study and public relations. For example, further—they are noticed, which suggest original levers of influence on stability in the highest education, which can be used to accelerate stable development. It is this synergy that characterizes the work of a "sustainable university" versus "greening a campus" or as a stand-alone work (Beringer and Adomßent 2008).

2 Discussion

Transforming a campus towards sustainability is a big challenge. And this is a step towards sustainability that not all higher education is capable of.

A university's sustainability goals should be both subsystems at the same time. To work on the implementation of the foundations and practices of resilience in both subsystems at the same time, sustainable institutional work recognizes the importance of top-down alignments, these as a declaration of institutional resilience, still ascending initiatives in the form of separate campaigns, initiatives and plans, these as a campaign to reduce energy consumption or the introduction of an environmental management system (Albrecht et al. 2007).

Institution as an organization and access to power, which students are deprived of because of their role and position in the academy, give students the project with goals it might help to increase awareness of global changing and increase understanding of sustainability and greening campuses (Beringer and Adomßent 2008). The abilities of students who make the campus a living world of sustainable development ideas of landscaping, hobbies towards greening, contributions are considered assets in sustainable institution projects.

House considered green if it can help save on disposal costs. If you invest in energy, this will save on electricity costs. Campus Management can be supported by technical staff who are aware of the need for a green campus, with the help of students and educators. Some green campus strategies include: conserving energy in buildings, waste management, air consumption, biodiversity on campus land, and transportation. Using environmental design concepts on campus because it has the potential to improve the quality of life for students as users. Several characteristics which are in homes can guarantee the comfort of use, for example, good natural lighting in stunning rooms, good ventilation, clean air, and a campus with a number of plants (Tamiami et al. 2018).

By its physiological structure, the campus is part of the city, which means that parking lots, green spaces, and buildings are considered to be the leading urban substances, elements of campuses. The architecture of new educational institutions is designed as a network of three-dimensional connections between teachers, students, and researchers for the dissemination, production, and exchange of knowledge inside and outside the building. Almost all new institutes are greening their campus, constructing new low-energy buildings to obtain environmental certifications such as LEED, HQE, WELL, Fitwel, etc. However, some of the certifications not implementing energy savings of renovating existing homes which at the same time contributes to their historical and cultural significance.

The space types on campus that were provided for better function in university are education and research, residential, retail and leisure, related business, and infrastructure (den Heijer and Curvelo Magdaniel 2012). These types provide the knowledge base, quality of life, economic base, and associability to university (Heijer and Curvelo Magdaniel 2012). The benefits of green campuses are that students can come with new ideas towards greening a campus and create an environment that is aware of environmental problems.

The most know rating certification LEED offers LEED Campus Certification, which certifies multiple projects in the same area, such as a campus or commercial development. LEED Campus Certification is not a ranking system it is a guide to forms and requirements for educational buildings and areas.

However, sustainable development instruments such as the Sustainability Assessment and Tracking System (STARS 2010) are composed of 67 indicators in three main categories. It is a clear assessment tool. Given that it is impossible to determine the existence of sustainability of higher education and a clear theoretical basis, this approach has become one of the most popular tools. The Pennsylvania State Performance Report's approach shows a different approach and covers 10 different topics, including energy, water, materials, food, and land, transportation, built environment, community, research, and decision making. Community and governance indicators are weak in this approach and lack indicators for diversity and services. However, this approach is being put into practice through the participation of a group of 30 students and several professionals.

Some certifications focus on sustainable construction and sustainable building design. However, after the completion of the project, the assessment does not evaluate the building through the use of the building. There is a possibility that residents in a certified building use more energy or water than residents in other buildings. The project's efforts to reduce electricity and water consumption do not include a draft on how the state's project introduces draft text in the draft text.

3 Conclusion

A sustainable campus means an institution that personally contributes to building a resilient society through community outreach and campus construction, education, research, etc.

A sustainable campus is not just a low-impact campus, it aims to provide practical and multifaceted assistance for the prosperousness of society by expanding education and research-based on public issues as a university-wide policy and understanding the creation of institutions in harmony with the surrounding lands.

Sustainability, as the part of the architecture, can be expressed by the introduction of the green building however, the introduction must be with knowledge as everything sustainable cannot be green.

If you focus on green building and choose the method, provide an opportunity for the university to become competent in the ecological, social, and economic interaction of people and surrounding systems and introduce this into the planning system of universities, then the implementation of university practice will reduce the consumption of energy, materials, and emissions that pollute the environment.

Using examples from other universities and finding a suitable method for each campus, that's the way of possibility to choose the correct proportions of the green building method and sustainability in higher education campuses.

References

Albrecht P, Burandt S, Schaltegger S (2007) Do sustainability projects stimulate organizational learning in universities? Int J Sustain High Educ 8(4):403–415

Beringer A, Adomßent M (2008) Sustainable university research and development: Inspecting sustainability in higher education research. Environ Educ Res 14(6):607–623. https://doi.org/10.1080/13504620802464866

den Heijer A.C, Curvelo Magdaniel FTJ (2012) The university campus as a knowledge city: exploring models and strategic choices. Int J Knowl-Based Dev 3(3):283–304

Federal Office for Spatial Development ARE (n.d.) 1987: Brundtland Report. Bundesamt für Raumentwicklung ARE. https://www.are.admin.ch/are/en/home/media/publications/sustainable-development/brundtland-report.html. Accessed 29 Sep 2021

Federal Office for Spatial Development ARE (n.d.) 2030 sustainable development strategy. Bundesamt für Raumentwicklung ARE. https://www.are.admin.ch/are/en/home/media/publications/sustainable-development/strategie-nachhaltige-entwicklung-2030.html. Accessed 5 Oct 2021

Finlay J, Massey J (2012) Eco-campus: applying the ecocity model to develop Green University and college campuses. Int J Sustain High Educ 13(2):150–165. https://doi.org/10.1108/14676371211211836

Gough S, Scott W (2007) Higher education and sustainable development: paradox and possibilities. Routledge, London

Hokkaido University Sustainable Campus Management Office. (n.d.). What is a sustainable campus? Hokkaido University Sustainable Campus Management Office. https://www.osc.hokudai.ac.jp/en/what-sc.

How BREEAM Certification Works. (5 September 2019). https://www.breeam.com/discover/how-breeam-certification-works/. Accessed 18 June 2020

Nine Categories of LEED Points. (31 January 2017). https://banyanwater.com/9-categories-of-leed-points/. Accessed 18 June 2020

World Green Building Council. (n.d.). Our Story. https://www.worldgbc.org/our-story. Accessed 18 Sep 2020

STARS, Sustainability Tracking Assessment & Rating System. The Sustainability Tracking, Assessment & Rating System. (30 April 2020). https://stars.aashe.org/

Tamiami Fachrudin H (2020) Green campus concept based on architect perspective. IOP Conf Ser Mater Sci Eng 801:012028. https://doi.org/10.1088/1757-899x/801/1/012028

Tamiami H, Khaira F, Fachrudin A (2018) Green design application on campus to enhance student's quality of life. IOP Conf Ser Mater Sci Eng 309:012022. https://doi.org/10.1088/1757-899x/309/1/012022

Thomashow M (2014) The nine elements of a sustainable campus. Sustainability. J Record 7(3):174–175. https://doi.org/10.1089/sus.2014.9788

Green Campus Improvement: Using Green Building Rating Systems in Universities

Aizhan Syidanova, Çiğdem Çağnan, and Dilber Uzun Ozsahin

Abstract Education plays a significant role in transforming society because it has become a strategy for promoting sustainable development. Educational and research work can lead to changes in the behavior and lifestyles of society, and it is also relevant for the dissemination of knowledge. Eco-friendly architectural ideas combined with sustainability and energy efficiency. Campuses lead to the principle of independent existence, as well as to the consumption of natural resources.

The process of organizing green university campuses is standout from the already established practice of the institutions of the world, however with global green strategies and global sustainability goals, achieving environmental sustainability on campuses may change.

Keywords Green campus · History of green campuses · Environmental health · Sustainability

1 Introduction

Campus (English campus)—a university campus, including, as a rule, classrooms, research institutes, living quarters for students, libraries, classrooms, canteens, etc. The word "campus" is of Latin origin means a field and open space. For the first time,

A. Syidanova (✉) · Ç. Çağnan
Department of Architecture, Near East University, Nicosia, TRNC, Turkey

Ç. Çağnan
e-mail: cigdem.cagnan@neu.edu.tr

D. U. Ozsahin
Medical Diagnostic Imaging Department, University of Sharjah, College of Health Science, Sharjah, United Arab Emirates
e-mail: dozsahin@sharjah.ac.ae

Department of Biomedical Engineering, Near East University, P.O. Box: 99138, Nicosia, TRNC, Mersin 10, Turkey

Center of Operational Research in Healthcare, Near East University, P.O. Box: 99138, Nicosia, TRNC, Mersin 10, Turkey

D. Uzun Ozsahin et al. (eds.), *Decision Analysis Applied to the Field of Environmental Health*, Professional Practice in Earth Sciences,
https://doi.org/10.1007/978-3-030-96682-9_6

the "campus" has named the Territory of Princeton University in the XVIII century (Diner 2017).

In the XX century, the concept of "campus" became broader because all institution property began to be included (Diner 2017). The campus is a term referring mainly to educational institutions in the world, where the presence of a university campus is considered the emblem of the institution, one of the characteristics of its prestige. Buildings for traditional campuses have formed in Europe but later spread throughout the globe. Firstly its was practical uses for university, for creating space for education. Later, it became an addition to prestige, a necessity for attracting students in a large number. The needs enough expanse for the student from different parts of the world. The space for creating all the conditions for scientific work. In this way, talk about campus space as a justified the need to establish an institute, then the transition to a "campus" system is the real and future of institutes in the world of science (Diner 2017).

In 1987 the UN World Commission on Environment and Development started to realize a concept called is Sustainable Development, 1987, Brundtland Report (Federal Office for Spatial Development ARE 2021). Advances in integrating sustainable development issues into university governance have intensified, especially in America and Europe; 1990s–2000s, were made statements to educational institutions to promote sustainability.

Recognizing that the architectural environment in general, and buildings in particular, play a role in the negative impact of humans on the natural environment, a declaration of Interdependence for a Sustainable Future was adopted at the 1993 World Congress of Architects (Declaration of Interdependence for a Sustainable Future 2006). In this way, architects have every chance to have an impact on restoring the ecological balance and ensuring the highest quality of life for the population, creating a building environment that meets human needs and at the same time protects or also improves the natural environment. The architectural environment and sustainability have been termed "sustainable architecture" (Declaration of Interdependence for a Sustainable Future 2006). Despite the lack of a clear scientific definition of sustainable architecture, there is a common understanding of its guiding principles and the recognition that following these principles requires a radical change in all design and construction practices.

To obtain clean breathing air means that sustainability is prioritized afterward in reduced air pollution and environmental toxins. The World Health Organization (WHO) has published a study that shows that air pollution causes about 4.2 million deaths annually worldwide. Today the word "sustainability" is widely used and can be included in various aspects of people's lives. Improving the university is a common question. Currently, assessment tools with rating systems are being used to find the best building management system for higher education and its campus. So the question of the work what should improve university for the healthy environment inside of the campus.

The global world governance system World Green Building Council (WGBC) is gaining popularity. By 2016, 1.04 billion square meters of green development had been certified worldwide through green building membership councils (WGBC).

WGBC recognizes the power, the tools used in the ranking, support their use. We understand that all rating instruments are individual and appropriate for their particular market. The GBC takes a neutral approach to assessment tools and does not recommend one assessment tool over another.

GBC believes that every green building rating tool meets quality standards. In 2015, WGBC published Quality Assurance Guidelines for Green Building Rating Instruments—a step-by-step guide for operators of new, emerging, and related standards.

Countries established joint councils of higher education institutions; in particular, the Environmental Association of Universities and Colleges (EAUC) of the United Kingdom and Ireland in 1996, and 2006 the USA and Canada ISCN (International Sustainable Campus Network) and AASHE (Association for the Advancement of Sustainability in Higher Education) and in 2013 in Japan the Office of Sustainable Campus Management of Hokkaido University (Tumbas et al. 2015). The beginners in the greening universities of these countries became the pioneers of the Sustainable Campus program.

Following the U.S. Environmental Protection Agency, resilience builds and supports an environment in whichever people and nature are likely you will be present in a productive agreement to satisfy the social, financial, and other needs of today and future generations.

The significance of resilience lies in the foundations of the future and is more rapidly standardized than those used to define the greenish structure.

Sustainable products reduce your environmental impact by providing products from reliable sources that are either fully renewable or environmentally friendly. Green products are goods that do not affect the surrounding land, do not pollute the air, and do not affect the long-term supply (Martty 2015).

Green building is defined by the creation of structures and the use of environmental and resource-saving processes throughout the life of a building, from placement to design, construction, operation, maintenance, renovation, detailed design, and even demolition of a building. The system expands and accompaniment the basic nuances of house design associated with efficiency, durability, and comfort benefit. Greenish construction is used as an environmentally friendly or highly efficient building (US Environmental Protection Agency).

United States Environmental Protection Agency (US EPA) commented that modern green building has emerged from environmental requirements and implementation of an energy-efficient and environmentally friendly construction method. In the 1970s, the rise in oil prices prompted research and energy efficiency activities to look for renewable energy (US Environmental Protection Agency). Together with the ecological movement of the 1960s and 1970s spurred early experiences with advanced greenish homes (US Environmental Protection Agency).

2 Discussion

Green building is a definition used to describe how to make the environment healthier. The main concerns are energy and resource efficiency, air pollution, soil and water well-being. Green building as a subject shows how to maintain the building environment, defines a way to create a sustainable environment, and becomes a healthy building using rating standards.

XXI century or Now as practice shows, some institutions in the world are already switching over to the system of organizing college campuses. New institutes are greening their campus, constructing new low-energy buildings to obtain environmental certifications. Campuses around the world are starting to incorporate sustainable campus principles into plans.

Campus greening strategies are required to connect green curricula, greenish procurement policymaker, energy conservation, air, recycling, transportation, planning and design, greenish classrooms, greenish labs, greenish IT and learning, teaching, and learning (Tamiami Fachrudin 2020).

A modern campus is not only a complex of buildings but also educational projects for which the educational space changes. A development program is required, which provides for the transformation of the architecture and spatial structure of the landscape, these ideas are manifest in the planning and functional zoning, as well as in the architectural appearance of buildings and structures.

Green University targets to research the environmental issues, innovation and uses them in the day-to-day management of the campus. A green university has a sustainability rule built into the university's operational and educational procedures. The benefits of campus landscaping include student participation, visibility, nature display, economic acquisition, documentation, curriculum interconnection, and student staff mobilization (Tamiami Fachrudin 2020).

The description "stability" has a definition that derives from the term sustainable agriculture, which means that any plant or animal product produced using agricultural methods that protect the environment, the well-being of the population, human society, and animals. Prosperity is the ability of future generations to do the same, which is sustainability.

Green campuses for universities use the seven principles from an architecture point of view: building, water conservation, energy consumption, landscape, waste management, transportation, and education (Tamiami Fachrudin 2020).

People associate sustainability with high financial costs, and money will be saved in the subsequent use of the building. Although, investing in the future may require upfront costs to reduce energy consumption through renewable energy sources.

The environment gives the right to win when people make choices towards organic farming, responsible land use, and habitat conservation. When it comes to organic, farming reduces greenhouse gas emissions and air pollution. Likewise, preserving and planting trees means better air quality.

The introduction of environmentally friendly materials and resources with low energy content and the impact on the environment is considered the most important substances of a durable building, as well as the introduction of water in appliances, faucets, and showerheads, the reintroduction of wastewater, and the reintroduction of rainwater for landscaping, and other non-drinking purposes (Martty 2015).

3 Conclusion

According to the structure of the campuses, the institutes are obliged to guarantee the best living and working conditions. Apart from this, institutions are obliged to fulfill the public and environmental promises arising from their existence as social institutions. Universities are more concerned with sustainability, which prompts them to start green programs.

Because institutions make an active contribution to the life of society, they are obliged to take part in ensuring the sustainability of their ecological development. Dissemination of knowledge and research on sustainable development, as well as information on how universities are using environmental education programs.

References

Declaration of interdependence for a sustainable future (n.d.) http://www.comarchitect.org/wp-content/architectsguide/declaration_of_interdependence_for_a_sustainable_future.htm. Accessed 30 Sept 2021

Environmental Protection Agency (n.d.). EPA. https://archive.epa.gov/greenbuilding/web/html/about.html#:~:text=The%20contemporary%20green%20building%20movement,and%20find%20renewable%20energy%20sources.

Federal Office for Spatial Development ARE (n.d.) 1987: Brundtland Report. Bundesamt für Raumentwicklung ARE. https://www.are.admin.ch/are/en/home/media/publications/sustainable-development/brundtland-report.html. Accessed 29 Sept 2021

Golden Arrow (n.d.) Why are green initiatives important? https://www.goldenarrow.com/blog/why-are-green-initiatives-important. Accessed 3 Oct 2021

Tamiami Fachrudin H (2020). Green campus concept based on architect perspective. IOP Conf Ser Mater Sci Eng 801: 012028. https://doi.org/10.1088/1757-899x/801/1/012028

Tamiami H, Khaira F, Fachrudin A (2018) Green design application on campus to enhance student's quality of life. IOP Conf Ser Mater Sci Eng 309: 012022. https://doi.org/10.1088/1757-899x/309/1/012022

Tumbas P, Matkovic P, Sakal M, Pavlićević V (2015) Sustainable university: assessment tools, factors, measures and model. In: 7th international conference on education and new learning technologies (EDULEARN 2015), Barcelona - Spain

Martty M (2015) The difference between green and sustainable – architecture. construction engineering property. Architecture Construction Engineering Property. https://sourceable.net/difference-green-sustainable/

Diner SJ (2017) College campus history: schools took students out of cities. Time. https://time.com/4751301/universities-colleges-history/

Vierra S (2019) Green building standards and certification systems. WBDG. https://www.wbdg.org/resources/green-building-standards-and-certification-systems

WHO (n.d.) Air pollution. https://www.who.int/health-topics/air-pollution#tab=tab_1. Accessed 3 Oct 2021
WorldGBC (n.d.) https://www.worldgbc.org/. Accessed 03 Oct 2021

Environmental Impact Assessment for the Production of Aggregates Used in the Construction Industry by Using MCDA

Mustafa Alas, Dilber Uzun Ozsahin, Huseyin Gokcekus, Berna Uzun, and Shaban Ismail Albrka

Abstract The need for addressing the environmental issues related to the production of aggregates for use in construction purposes cannot be ignored especially considering the amount of aggregates needed particularly to produce asphaltic and cemented concrete to meet the requirements for infrastructure construction to accommodate societal demand. Life cycle assessment (LCA) methods have been an efficient tool to evaluate the environmental burdens of manufacturing aggregates. Mostly, the LCA results have been interpreted individually by considering certain criteria in the environmental impact assessment. Herein, a multi criteria decision analysis (MCDA) was employed by using the Fuzzy-PROMETHEE technique to evaluate the impact of producing aggregates to the environment based on multiple criteria. According to the LCA results, manufacturing of different types of aggregates has certain positive and negative aspects compared to other alternatives in terms of environmental impact and the direction of environmental effect have varied based on the particular criteria. The MCDA study conducted in this study assisted in ranking the most preferable sources of aggregates and their manufacturing process by considering the complete

M. Alas · H. Gokcekus · S. I. Albrka
Faculty of Civil and Environmental Engineering, Department of Civil Engineering, Near East University, P.O. Box: 99138, Nicosia, TRNC, Mersin 10, Turkey

D. U. Ozsahin
Medical Diagnostic Imaging Department, University of Sharjah, College of Health Science, Sharjah, United Arab Emirates
e-mail: dozsahin@sharjah.ac.ae

Faculty of Engineering, Department of Biomedical Engineering, Near East University, P.O. Box: 99138, Nicosia, TRNC, Mersin 10, Turkey

D. U. Ozsahin · B. Uzun (✉)
Center of Operational Research in Healthcare, Near East University, P.O. Box: 99138, Nicosia, TRNC, Mersin 10, Turkey
e-mail: berna.uzun@neu.edu.tr

B. Uzun
Faculty of Arts and Sciences, Department of Mathematics, Near East University, P.O. Box: 99138, Nicosia, TRNC, Mersin 10, Turkey

Department of Statistics, Carlos III University of Madrid, 28903 Getafe, Spain

D. Uzun Ozsahin et al. (eds.), *Decision Analysis Applied to the Field of Environmental Health*, Professional Practice in Earth Sciences,
https://doi.org/10.1007/978-3-030-96682-9_7

set of environmental impact indicators. Based on the Fuzzy-PROMETHEE results it was observed that, the manufacturing of recycled aggregates has the least environmental impact while the highest environmental impact was observed for the marine aggregates production.

Keywords Life cycle assessment · Multiple criteria decision analysis · Fuzzy-PROMETHEE · Mineral aggregates · Environmental impact assessment

1 Introduction

Aggregates are foremost the most required minerals to produce cementitious mixtures for building construction, bituminous mixtures for pavement construction and for use in various other structural constructions. Aggregates typically constitutes 60–75% of cementitious mixtures and around 93–95% of bituminous mixtures consist of course and fine aggregates (Habibi and Ghomashi 2018; Ameri 2018). According to Silva et al. (2021) in 2020, the global consumption of aggregates for the construction industry was approximately 46 billion metric tons and as a future forecast, it is expected to increase every year by 2.5% particularly due to infrastructure requirements in underdeveloped and developing countries. Common sources which the aggregates are extracted from include primary sources; ingenious rocks, sedimentary rocks, sand and gravel deposits and secondary sources; recycled unbound inert waste, recycled asphalt and recycled concrete (Knepper et al. 1995; Mendoza et al. 2017). In UK which is one of the leading countries for using secondary sources for producing materials of construction, around 28 to 37% of raw materials are derived from secondary sources and recycled aggregates to meet the overall demand of aggregates in the construction industry (Herbst et al. 2016). Since aggregates are an essential part of a mixture design and that, they are used in significant amount whether for a cementitious or a bituminous mix, there have been a plethora of research focusing on improving the performance of mix design based on utilizing different aggregate types, recycled aggregate options and their characteristic properties (Institute 2014). On the other hand, it is known that, the aggregate selection for a mix design is often limited by the locally available sources and also by the availability of the technology to obtain or to mine the mineral aggregates (Wilburn and Goonan 1998).

Although engineers are mostly concerned with the technical aspects of aggregates to achieving better strength and durability for construction purposes, there is a gap in the knowledge regarding to the environmental impact of mining, processing and utilizing mineral aggregates in the production of construction materials (Schiappacasse et al. 2020). To shed a light on this matter, researchers have previously utilized life cycle assessment (LCA) methods by using existing and/or by developing new life cycle inventories (LCI) to evaluating the impact of materials, processes and activities related with the construction works (Gulotta et al. 2018). LCA is proven to be an effective tool to evaluating the environmental burdens of various construction applications which can be used to assist in decision making and strategic planning.

The primary purpose of utilizing LCA has been to assess the environmental friendliness of materials and processes by considering different assessment criteria such as greenhouse gas emissions, energy consumption, global warming potential, ecotoxicity and ozone layer depletion (Park et al. 2019; Sackey et al. 2019; Ferreira et al. 2016).

As acknowledged in the literature, a number of high impact research considered cradle to gate LCA while others were specifically targeted to certain phases of construction works such as extracting, manufacturing, transporting and recycling and disposal of the construction materials (Park et al. 2019; Kleijer et al. 2017). A number of earlier works regarding to LCA was attributed to the environmental impact of aggregates production for the manufacture of construction concrete and asphaltic concrete (Korre and Durucan 2009). Kleijer et al. (2017) studied the environmental impact of ordinary and recycled concrete based on product specific life cycle assessment. Their findings indicated that, if the transportation distance for the recycled materials to the construction site is less than 25 km, recycled concrete option only performs slightly better than the ordinary concrete in terms of detrimental environmental effects. (Hossain et al. 2016) conducted LCA study by making a comparison between aggregates from natural sources, waste materials from construction and demolition works and from recycled waste glass. They concluded that, by utilizing waste obtained from construction and demolition and waste glass, a remarkable reduction on environmental impact can be achieved. Ferreira et al. (2016) evaluated the incorporation of steel slag as aggregates for bituminous pavements and its potential impact to the environment. They have reported that, along with the technical improvements in the performance characteristics of the bituminous mixes, steel slag utilization as aggregates was a viable solution to reduce the environmental impact particularly in terms of ozone layer depletion. According to Praticò et al. (2020), materials production composes more than 60–70% of the total environmental effect in the asphalt concrete pavement construction. The results of their study showed that, utilizing warm mix asphalt and recycled aggregates significantly reduces energy consumption, carbon footprint and a number of other major impact indicators.

LCA studies are able to reveal multiple environmental impact indicators at once. These indicators vary according to the assessment method and criteria based on the purpose of the study. Most of the LCA studies found in the literature were found to present their findings by considering different impact assessment criteria individually (Heede and Belie 2012). However, a multiple criteria assessment method that considers all impact indicators in one model is thought to provide a more complete assessment towards understanding the environmental burdens of materials and mix design choices to enabling better decision making process for designers and engineers (Silgado et al. 2018). Multiple Criteria Decision Analysis (MCDA) is a method that enables decision makers to rank, select and compare alternative options specifically when multiple conflicting criteria exists (Hermann et al. 2007). There have been numerous research available in the literature which have implemented MCDA to evaluate the structural, economic and environmental performance of construction materials and processes based on different multiple assessment criteria (Silgado et al. 2018). Also noted in the literature, several research have utilized LCA and MCDA in combination to make comparison between different materials and processes that are involved in construction industry. Some of the significant studies that have adopted LCA and MCDA methods in combination include (Hermann et al. 2007; Jato-Espino et al. 2014; Kim et al. 2013; Tošić et al. 2015; Vučijak et al. 2016). In this research, LCA analysis for the production of four different aggregate types were obtained from a report prepared by Waste and Resources Action Programme (WRAP) (Korre and Durucan 2009) and MCDA study was conducted to compare the effects of producing aggregates on the environment.

2 Data Acquisition

Data for the present study was collected from the case study reports conducted in UK during the Waste and Resources Action Programme. The readers are referred to the full report titled "EVA025: Aggregates Industry Life Cycle Assessment Model: Modelling Tools and Case Studies" for further review (Korre and Durucan 2009). The primarily objectives of the programme was to develop a lifecycle inventory (LCI) and to perform LCA study for the production of different type of aggregates that are used in the construction industry in UK. The product systems and boundaries included in their report were; primary aggregate systems; land-won aggregate systems (LW), crushed rock aggregate systems (CR), marine aggregate systems (MA) and secondary aggregate systems; recycled aggregate (RA) systems. LCI analysis involved resource inputs such as materials and energy consumption, wastes such as overburden waste and fines and emission streams which involved all the gaseous emissions such as CO_2, NO_x, SO_2 etc. The environmental impact assessment criteria included in the LCA were; Global warming potential, eutrophication, acidification, photo-oxidant formation, human toxicity, freshwater aquatic ecotoxicity, marine aquatic ecotoxicity, terrestrial ecotoxicity and ozone layer depletion. The phases of aggregate production systems that were considered in the LCA study included three different procedures; materials extraction, processing and waste management/restoration which were as demonstrated in Table 1. A number of key assumptions and limitations regarding the evaluation of impact categories were involved in the preparation of the report. Product distribution systems such as transport by road, rail, shipping and aggregate handling in transit were inclusive of the report while, certain phases in the aggregate production systems such as the manufacture of the capital equipment required for producing aggregates and the labour work were exclusive due to difficulties in allocation, drawing boundaries and interpreting data in the LCA study and therefore they were listed as limitations of the study.

Considering the limitations, inclusive and exclusive data for the production of four different aggregate types, environmental impact indicators were assessed according to nine different impact categories and the findings from the case studies reported in EVA025 were summarised in Table 2.

Table 1 Aggregates production systems

		Extraction	Processing	Restoration
Primary aggregates	Crushed rock aggregates	Overburden removal	Primary crushing	Waste landfilling
		Primary fragmentation	Scalping screening	Site preparation for restoration
		Loading and hauling	Secondary crushing	Re-vegetation
			Tertiary crushing	Re-instalment
			Quaternary crushing	
			Final screening	
	Land-won aggregates	Excavation	Pre-processing storage	Waste landfilling
			Scalping screening	Site preparation for restoration
			Crushing	Re-vegetation
			Size screening	Re-instalment
			Washing-scrubbing	
			Wet classification	
			Dewatering	
			Grinding	
			Product storage	
	Marine aggregates	Marine aggregates loading	Pre-processing storage	
		Marine aggregates discharge	Scalping screening	
			Crushing	
			Size screening	
			Washing-scrubbing	
			Wet classification	
			Dewatering	
			Grinding	
			Product storage	
Secondary aggregates	Recycled aggregates		Waste reception	
			Pre-screening	
			Screening	
			Crushing	
			Conveying and magnetic separation	
			Washing-scrubbing	
			Secondary screening	
			Secondary crushing	
			Material transport and storage	

Table 2 Summary of the impact assessment results per unit process per one tonne of aggregate produced

	Global Warming	Eutrophication	Acidification	Photo-oxidant formation	Human toxicity	Freshwater aquatic ecotoxicity	Marine aquatic ecotoxicity	Terrestrial Ecotoxicity	Ozone layer depletion
	kg CO_2 eq.	kg PO_4 eq.	kg SO_2 eq.	kg ethylene eq.	kg 1,4-DB eq.	kg 1,4-DB eq.	kg 1,4-DB eq.	kg 1,4-DB eq.	kg R11 eq.
CR	2.0000000	0.0007145	0.0116900	0.0008360	0.3725000	0.0074900	161.5700000	0.0034050	0.0000003
RA	2.4200000	0.0007060	0.0121300	0.0008000	0.1733000	0.0019550	30.4500000	0.0008620	0.0006726
LW	1.3300000	0.0004545	0.0074200	0.0043155	0.2370000	0.0040400	76.5970000	0.0018395	0.0000002
MA	37.8550000	0.1041000	0.6765000	0.0539000	9.8600000	4.5750000	325.1000000	0.0251000	0.0000001

3 Multiple Criteria Decision Analysis (MCDA)

The analysis of experimental results, findings from the literature and evaluation of results from the field surveys are only useful to engineers if an efficient decision making process can be conducted. MCDA is an efficient method that assists researchers and engineers to take decisions based on available information particularly when multiple criteria are required to be taken into account. Although a number of different MCDA techniques such as Analytic Hierarchy Process (AHP) and Technique for Order of Preference by Similarity to Ideal Solution (TOPSIS) are available, in this study Fuzzy-PROMETHEE technique was utilized due to its simplistic nature of solving complex engineering problems by considering different assessment criteria. In the present study, the materials assessed and the criteria used to assessing the potential environmental impact of the manufacturing process for the aforementioned materials were as expressed in Table 2. Another advantage for the Fuzzy-PROMETHEE method was that, the data points were firstly fuzzified rather than using crispy data points which includes Boolean logic and can bias the analysis results. For developing the Fuzzy-PROMETHEE model, each criteria were assigned equal weight while comparing the alternative types of aggregates. In making comparison between different alternatives, a preference function was assigned for each criteria separately. A preference function enables the user to set numerical boundaries as to when an alternative should be prioritized over another. Herein, the Gaussian preference function was utilized. In the next step, the outranking relation between the alternatives was determined and the positive and negative outranking were compared partially and the partial pre-order on the alternatives were defined. Finally, the net flows for each alternatives were computed as the difference between the positive and negative outranking. A higher net flow indicated that, an alternative is superior over the others and a lower net flow indicated the versa.

4 Results and Discussion

The results obtained from the LCA study were based on the environmental impact of manufacturing primary aggregates and secondary aggregates considering the extraction, processing and restoration processes. From Table 2, it can be observed that, each alternative type of aggregate have pros and cons over the other alternatives regarding to the different criteria used in the evaluation of environmental impact of manufacturing the aggregates. The manufacture of land-won aggregates was observed to have the lowest impact on the environment in terms of eutrophication, acidification, photo-oxidant formation and global warming potential while, the environmental impact of manufacturing land-won aggregates were significantly higher regarding to the other criteria. On the other hand, the production of recycled aggregates were found to have the least impact on the environment regarding the human toxicity, freshwater, marine and terrestrial ecotoxicity. Besides, although for all criteria the production of marine aggregates have shown to have the highest contribution to environmental hazard, in terms of ozone layer depletion marine aggregates were observed to have the lowest impact to environment in this criteria. Based on these results, a dilemma raised due to conflict in determining which alternative is more preferable over the other since each alternative have shown distinct potential to contribution to environmental impact according to different criteria. Therefore, the Fuzzy-PROMETHEE technique was applied to provide a comprehensive multi criteria decision analysis (MCDA) to determining the best alternative that has the least impact on the environment by considering all of the criteria at once. The results from the MCDA were illustrated in Fig. 1. It can be seen that, recycled aggregates, land-won aggregates, and crushed rock aggregates demonstrated positive ranking almost in all of the criteria while, regarding the ozone layer depletion and marine aquatic ecotoxicity, the production of recycled aggregates and crushed rock aggregates possessed negative ranking respectively. On the other hand, Marine sand and gravel aggregates manufacturing have shown negative outranking in all criteria except for the ozone layer depletion.

The results of the analysis also revealed the positive and negative outranking flow for each alternative types of aggregates and the computed net outranking flow. Demonstrated in Table 3, each alternative was ranked based on the Fuzzy-

Fig. 1 Fuzzy-PROMETHEE evaluation results

Table 3 Complete ranking for aggregates from different sources based on environmental impact

Rank	Alternative	Positive outranking flow	Negative outranking flow	Net flow
1	RA	0.2114	0.0027	0.2087
2	LW	0.1776	0.0370	0.1406
3	CR	0.1376	0.0625	0.0625
4	MA	0.0000	0.4118	−0.4118

PROMETHEE outcomes. Recycled aggregates were found to yield the highest positive and the lowest negative outranking flow and presented the highest net outranking flow indicating that, considering the combination of all criteria, the production of recycled aggregates have the lowest environmental impact. The land-won aggregates, crushed rock aggregates and marine sand and gravel aggregates were consecutively ranked after the recycled aggregates while only manufacturing of marine sand and gravel aggregates resulted in negative net outranking flow. The outcomes observed herein, have been verified and contradicted in a number of similar studies. The findings from a study conducted by Rosado et al. (2017) have reported that, the manufacture of recycled aggregates is favourable compared to the manufacture of aggregates from natural resources and mix production of aggregates using both primary and secondary sources of raw materials in terms of respiratory inorganics, terrestrial ecotoxicity, land occupation, global warming and non-renewable energy aspects. However, in another study conducted by Park et al. (2019) noted that, the use of recycled aggregates resulted in double the amount of environmental impact caused by producing aggregates from natural sources mainly, due to simplicity and less energy consumption for producing the aggregates from natural sources. On the other hand, their findings indicated that, recycled aggregates were far more a better alternative in terms of sustainability compared to natural aggregates because using recycled aggregates foreclose the depletion of natural resources. Never the less in each study different LCA methods and assessment criteria were used and it is difficult to compare the findings from different LCA studies.

5 Conclusion

The acquired data from the case studies for the production of different types of aggregates from primary and secondary sources were analysed by using the Fuzzy-PROMETHEE technique. The impact of manufacturing each type of aggregate to the environment were analysed by considering a total of nine different criteria. Based on the analysis results from the PROMETHEE, the manufacture of recycled aggregates were found to be the most environmental friendly process while aggregates produced from marine sand and gravel were found to have the highest contribution to detrimental environmental effect. A shortcoming for the present study can be noted as, only the manufacturing phase for the aggregates were considered in the life cycle assessment and therefore the findings herein can only be valid for this phase. As a future research suggestion, it can be recommended that, an MCDA can be applied based on acquired data from field experiments or data available from an LCA study in the literature which all phases of aggregates such as production, construction, in service and restoration are included. This way, the analysis results is thought to be more useful for engineers and decision makers.

References

Ameri M et al (2018) Investigating effects of nano/SBR polymer on rutting performance of binder and asphalt mixture. In: Advances in Materials Science and Engineering, vol 2018

Ferreira VJ et al (2016) Evaluation of the steel slag incorporation as coarse aggregate for road construction: technical requirements and environmental impact assessment. J Clean Prod 130:175–186

Gulotta T, Mistretta M, Praticò F (2018) Life cycle assessment of roads: material and process related energy savings

Habibi A, Ghomashi J (2018) Development of an optimum mix design method for self-compacting concrete based on experimental results. Constr Build Mater 168:113–123

Herbst A et al (2017) Benchmarking the EU reference scenario 2016: an alternative bottom-up analysis of long-term energy consumption in Europe. In: ECEEE Summer Study Proceedings

Hermann BG, Kroeze C, Jawjit W (2007) Assessing environmental performance by combining life cycle assessment, multi-criteria analysis and environmental performance indicators. J Clean Prod 15(18):1787–1796

Hossain MU et al (2016) Comparative environmental evaluation of aggregate production from recycled waste materials and virgin sources by LCA. Resour Conserv Recycl 109:67–77

Institute A (2014) MS-2 asphalt mix design methods. Lexington Kentucky, USA

Jato-Espino D et al (2014) A review of application of multi-criteria decision making methods in construction. Autom Constr 45:151–162

Kim S-H et al (2013) Environmental impact assessment and eco-friendly decision-making in civil structures. J Environ Manage 126:105–112

Kleijer A et al (2017) Product-specific Life Cycle Assessment of ready mix concrete: comparison between a recycled and an ordinary concrete. Resour Conserv Recycl 122:210–218

Knepper DH, Langer WH, Miller S (1995) A survey of natural aggregate properties and characteristics important in remote sensing and airborne geophysics. Nonrenew Resour 4(1):99–120

Korre A, Durucan S (2009) EVA025–Final Report: Aggregates Industry Life Cycle Assessment Model: Modelling Tools and Case Studies. Waste and Resources Action Programme

Mendoza FJC, Altabella JE, Izquierdo AG (2017) Application of inert wastes in the construction, operation and closure of landfills: calculation tool. Waste Manage 59:276–285

Park W-J et al (2019) Analysis of life cycle environmental impact of recycled aggregate. Appl Sci 9(5):1021

Praticò FG et al (2020) Energy and environmental life cycle assessment of sustainable pavement materials and technologies for urban roads. Sustainability 12(2):704

Rosado LP et al (2017) Life cycle assessment of natural and mixed recycled aggregate production in Brazil. J Clean Prod 151:634–642

Sackey S, Lee D-E, Kim B-S (2019) Life cycle assessment for the production phase of nano-silica-modified asphalt mixtures. Appl Sci 9(7):1315

Schiappacasse P et al (2020) Construction aggregates and environmental policy integration in a one-party state: the case of Hoa Binh, Vietnam. Sustainability 12(17):6890

Silgado SS et al (2018) Multi-criteria decision analysis to assess the environmental and economic performance of using recycled gypsum cement and recycled aggregate to produce concrete: the case of Catalonia (Spain). Resour Conserv Recycl 133:120–131

Silva FA et al (2021) Preliminary analysis of the use of construction waste to replace conventional aggregates in concrete. Buildings 11(3):81

Tošić N et al (2015) Multicriteria optimization of natural and recycled aggregate concrete for structural use. J Clean Prod 87:766–776

Van den Heede P, De Belie N (2012) Environmental impact and life cycle assessment (LCA) of traditional and 'green' concretes: literature review and theoretical calculations. Cement Concr Compos 34(4):431–442

Vučijak B, Kurtagić SM, Silajdžić I (2016) Multicriteria decision making in selecting best solid waste management scenario: a municipal case study from Bosnia and Herzegovina. J Clean Prod 130:166–174

Wilburn DR, Goonan TG (1998) Aggregates from natural and recycled sources. US Geol Surv Circular 1176:36

Analyzing the Relationship Between Covid-19 and Proportions of Vaccine & Mobility

Effect of Vaccine and Mobility on Covid-19

Bilgen Kaymakamzade, Evren Hincal, Nezihal Gokbulut, and Tamer Sanlidag

Abstract In this paper, Covid-19 is studied as an infectious disease. A mathematical model is constructed to see the effect of vaccine and mobility on the spread of the disease. Firstly, model is proposed, then analysis of the model is given with the basic reproduction number, R_0, of the disease. With the formula obtained for R_0, numerical simulations are made to see the decrease and increase in susceptible, infected, recovered, and died individuals by changing the proportion of vaccinated individuals and mobility. Lastly, conclusions are given and it is stated that vaccination is not enough without any restrictions.

Keywords Covid-19 · Mathematical modelling · Basic reproduction number · Transmission dynamics · Vaccine · Mobility

1 Introduction

Valuable contributions have been made to epidemics by mathematicians that impacts humanity throughout history for defining, modelling, and controlling epidemic diseases by estimating their behavior (Çetin et al. 2008). To control epidemics, vaccination of the population is one of the most popular and effective idea to prevent the infection of susceptible people since it provides an immunization for the population (Lahariya 2016). Mathematical modelling is an efficient way which tries to find the best way of vaccination that should be preferred in a case of epidemic. In modelling,

B. Kaymakamzade (✉) · E. Hincal · N. Gokbulut
Department of Mathematics, Near East University, Nicosia, TRNC, Turkey
e-mail: bilgen.kaymakamzade@neu.edu.tr

E. Hincal
e-mail: evren.hincal@neu.edu.tr

N. Gokbulut
e-mail: nezihal.gokbulut@neu.edu.tr

B. Kaymakamzade · E. Hincal · N. Gokbulut · T. Sanlidag
DESAM Institute, Near East University, Nicosia, TRNC, Turkey

D. Uzun Ozsahin et al. (eds.), *Decision Analysis Applied to the Field of Environmental Health*, Professional Practice in Earth Sciences,
https://doi.org/10.1007/978-3-030-96682-9_8

proportion of the vaccinated people in a population and weakening of the immunity provided by vaccine are the most important issues that should be taken into consideration.

There are many mathematical models that dealt with the epidemics in literature. Kansagra and Farley (2009), Tuite et al. (2010), Kaymakamzade et al. (2016) and Mathews et al. (2009) are some of them that examined the dynamics of diseases while (Chadla et al. 2015; Baba et al. 2018; Wang et al. 2015; Zhen et al. 2016; Saha et al. 2014) constructed models to see the effect of vaccine on diseases.

Nowadays, the whole world is struggling with the epidemic *Covid-19* occurred as a result of *SARS-CoV-2* infection. It is from the coronavirus family that attacks the human respiratory system and affect from mild to severe (WebMD 2020). In the whole world, many restrictions, lock-downs, and quarantine are applied for the control of the spread. Until the 1th of March, approximately 115 million people infected by the virus and more than 2,5 million people passed away as a result of infection (Worldometer 2021). Now, one of the most popular idea for controlling the spread of Covid-19 is the foundation of a vaccine. This is because very successful vaccines have been introduced in previous years for the epidemics like H1N1, measles, smallpox, etc. (Scherer and McLean 2002). The purpose of a vaccine is to provide the immune system recognizing the virus before it enters the body. So, with vaccine, decrease in infectiousness of the disease and eradication of the virus are expected (Britannica 2020).

In epidemiology, mathematical modelling intends to analyze the dynamics of the epidemic to see the infectiousness of the disease. For this, basic reproduction number, denoted by R_0, of the epidemic is determined which shows the infectiousness of the disease. If R_0 values are less than 1, it means that the spread of the disease is under control while $R_0 > 1$ defines an epidemic; an uncontrolled spread of disease. Beside this, mathematical modelling helps to determine the minimum amount of vaccine proportion for the population (Dauhoo et al. 2018).

The rest of the paper is organized as follows: In Sect. 2, analysis of the model is provided including model formulation, equilibrium points of the model, basic reproduction number of the model, and stability analysis. In Sect. 3, numerical simulations are given. Lastly, in Sect. 4, conclusions are presented.

2 Model Analysis

2.1 Model Formulation

A new mathematical model is proposed to monitor the transmission dynamics of the novel coronavirus, *Covid-19*. The total human population at time t, given by $N(t)$, is divided into sub-populations containing susceptible individuals, $S(t)$, vaccinated individuals $V(t)$, individuals with mild infection, $I_m(t)$, individuals with severe infection, $I_s(t)$, recovered individuals, $R(t)$, and Covid-19 caused died individuals,

Table 1 Descriptions of the variables that are used in the model

Variables	Description
N	Total population
S	Susceptible individuals
V	Vaccinated individuals
I_m	Infected individuals with mild infection
I_s	Infected individuals with severe infection
R	Recovered individuals
D	Covid-19 caused died individuals

$D(t)$, such that $N(t) = S(t) + V(t) + I_m(t) + I_s(t) + R(t) + D(t)$.

$$\begin{cases} \frac{dS}{dt} = \lambda - rS(t) - \beta_1 S(t) I_m(t) - \beta_2 S(t) I_s(t) - \mu S(t), \\ \frac{dV}{dt} = rS(t) - [k_1 I_m(t) + k_2 I_s(t)]V(t) - (1 - k_1 - k_2)V(t), \\ \frac{dI_m}{dt} = e^{-\mu\tau}(\beta_1 S(t-\tau) + k_1 V(t-\tau)) I_m(t-\tau) - \beta_3 I_m(t) - \gamma_1 I_m(t), \\ \frac{dI_s}{dt} = e^{-\mu\tau}(\beta_2 S(t-\tau) + k_2 V(t-\tau)) I_s(t-\tau) + \beta_3 I_m(t) - (\gamma_2 + d) I_s(t), \\ \frac{dR}{dt} = \gamma_1 I_m(t) + \gamma_2 I_s(t) + (1 - k_1 - k_2)V(t), \\ \frac{dD}{dt} = d I_s(t). \end{cases}$$

The linear system of ODE's is constructed which can be seen from the equation above. Descriptions of the variables and parameters used in the model are stated in Table 1 and Table 2.

The constructed model is nonnegative for every $t \geq 0$, with respect to the population, each of its parameters, and variables. Hence, it can be easily proved that for each nonnegative initial prerequisite, the state variables of the model are nonnegative. From the system, we have

$$\frac{dN}{dt} = \lambda - \mu(S + R).$$

This means that $\frac{dN}{dt} \leq \lambda$. Hence, $\lim\ supN \leq \lambda$.

Therefore, the feasible region for the model will be

$$\pi = \left\{(S, V, I_m, I_s, R, D) \in R_+^6 : S \geq 0, V \geq 0, I_m \geq 0, I_s \geq 0, R \geq 0, D \geq 0, N \leq \lambda\right\}.$$

The system can be reduced to

$$\begin{cases} \frac{dS}{dt} = \lambda - rS(t) - \beta_1 S(t) I_m(t) - \beta_2 S(t) I_s(t) - \mu S(t), \\ \frac{dV}{dt} = rS(t) - [k_1 I_m(t) + k_2 I_s(t)]V(t) - (1 - k_1 - k_2)V(t), \\ \frac{dI_m}{dt} = e^{-\mu\tau}(\beta_1 S(t-\tau) + k_1 V(t-\tau)) I_m(t-\tau) - \beta_3 I_m(t) - \gamma_1 I_m(t), \\ \frac{dI_s}{dt} = e^{-\mu\tau}(\beta_2 S(t-\tau) + k_2 V(t-\tau)) I_s(t-\tau) + \beta_3 I_m(t) - (\gamma_2 + d) I_s(t). \end{cases}$$

Table 2 Descriptions of the parameters that are used in the model

Parameters	Description
λ	Recruitment rate
r	Rate of susceptible individuals to be vaccinated
β_1	Rate of susceptible individuals to be mild infected
β_2	Rate of susceptible individuals to be severe infected
β_3	Rate of mild infected individuals to be severe infected
μ	Rate of natural death
k_1	Rate of vaccinated individuals to be mild infected
k_2	Rate of vaccinated individuals to be severe infected
γ_1	Rate of mild infected individuals to be recovered
γ_2	Rate of severe infected individuals to be recovered
d	Death rate of infected individuals
τ	Latent period

2.2 Equilibrium Points

In order to find disease free equilibrium (DFE), severe infected free equilibrium, and endemic equilibrium points, each equation in the reduced system is equated to zero. They are obtained as follows:

$$E_0 = \left(S_0, V_0, I_{m,0}, I_{s,0}\right) = \left(\frac{\lambda}{r+\mu}, \frac{r\lambda}{(r+\mu)(1-k_1-k_2)}, 0, 0\right),$$
$$E_1 = \left(S_1, V_1, I_{m,1}, I_{s,1}\right) = \left(\frac{\lambda}{r+\mu+\beta_1 I_{m,1}}, \frac{r\lambda}{(r+\mu+\beta_1 I_{m,1})(1-k_1-k_2+k_1 I_{m,1})}, I_{m,1}, 0\right),$$
$$E_2 = \left(S_2, V_2, I_{m,2}, I_{s,2}\right) = \left(\frac{\lambda}{r+\mu+\beta_2 I_{s,2}}, \frac{r\lambda}{(r+\mu+\beta_2 I_{s,2})(1-k_1-k_2+k_2 I_{s,2})}, 0, I_{s,2}\right).$$

Here, $I_{m,1}$ and $I_{s,2}$ are the solutions of the following second-order equations.

$$A_1 I_{m,1}^2 + B_1 I_{m,1} + C_1 = 0,$$
$$A_2 I_{s,2}^2 + B_2 I_{s,2} + C_2 = 0,$$

where

$$A_1 = e^{\mu\pi}(\gamma_1 + \beta_3)\beta_1 k_1,$$
$$B_1 = e^{\mu\pi}(\gamma_1 + \beta_3)[(r+\mu)k_1 + (1-k_1-k_2)\beta_1] - \beta_1 k_1 \lambda,$$
$$C_1 = e^{\mu\pi}(\gamma_1 + \beta_3)(1-k_1-k_2) - \lambda\beta_1(1-k_1-k_2) - \lambda r k_1,$$
$$A_2 = e^{\mu\pi}(\gamma_2 + d)\beta_2 k_2,$$
$$B_2 = e^{\mu\pi}(\gamma_2 + d)[(r+\mu)k_2 + (1-k_1-k_2)\beta_2] - \beta_2 k_2 \lambda,$$
$$C_2 = e^{\mu\pi}(\gamma_2 + d)(1-k_1-k_2) - \lambda\beta_2(1-k_1-k_2) - \lambda r k_2.$$

In order to get a meaningful mild and severe endemic equilibrium points (MEP, SEP), above equation should have a positive solution, i.e., C_1 and C_2 must be less than zero. Hence, MEP and SEP are exists if

$$\frac{\lambda e^{-\mu\pi}}{(\gamma_1+\beta_3)(r+\mu)}\left[\beta_1 + \frac{k_1 r}{1-k_1-k_2}\right] > 1$$

and

$$\frac{\lambda e^{-\mu\pi}}{(\gamma_2+d)(r+\mu)}\left[\beta_2 + \frac{k_2 r}{1-k_1-k_2}\right] > 1,$$

respectively.

2.3 *Basic Reproduction Number*

The basic reproduction number, is the number of secondary cases per primary infection and it is denoted by R_0. In this study, R_0 formula is computed by using the next generation matrix (NGM) method, which represents the number of secondary cases produced by an infected individual with Covid-19 infection throughout his/her entire period of infection in a fully susceptible population (Driessche and Watmough 2002; Musa et al. 2019; Diekmann et al. 1990).

From the system, followings are obtained according to the NGM method.

$$f = \begin{bmatrix} e^{-\mu\pi}(\beta_1 S I_m + k_1 V I_m) \\ e^{-\mu\pi}(\beta_2 S I_s + k_2 V I_s) \end{bmatrix},$$
$$v = \begin{bmatrix} \beta_3 I_m + \gamma_1 I_m \\ -\beta_3 I_m + (\gamma_2 + d) I_s \end{bmatrix}.$$

So,

$$F = \begin{bmatrix} e^{-\mu\pi}(\beta_1 S_0 + k_1 V_0) & 0 \\ 0 & e^{-\mu\pi}(\beta_2 S_0 + k_2 V_0) \end{bmatrix},$$
$$V = \begin{bmatrix} \beta_3 + \gamma_1 & 0 \\ -\beta_3 & \gamma_2 + d \end{bmatrix}.$$

Hence, basic reproduction number is the spectral radius (dominant eigenvalue) of FV^{-1}, i.e., $R_0 = \rho(FV^{-1})$ which is

$$R_0 = \{R_{0,1}, R_{0,2}\},$$

where,

$$R_{0,1} = \frac{\lambda e^{-\mu\pi}}{(\gamma_1 + \beta_3)(r + \mu)}\left[\beta_1 + \frac{k_1 r}{1 - k_1 - k_2}\right]$$

and

$$R_{0,2} = \frac{\lambda e^{-\mu\pi}}{(\gamma_2 + d)(r + \mu)}\left[\beta_2 + \frac{k_2 r}{1 - k_1 - k_2}\right].$$

2.4 Stability Analysis

In this section necessary theorems with proofs are given. For global stability analysis, Lyapunov function and Lyapunov-Laselle invariance principle are used.

Theorem 2.1. For the proposed model, the disease-free equilibrium is globally asymptotically stable whenever $R_0 < 1$.

Proof. Consider the Lyapunov function

$$\begin{aligned} V = {} & S_0 g\left(\frac{S}{S_0}\right) + V_0 g\left(\frac{V}{V_0}\right) + e^{\mu\tau} I_m + e^{\mu\tau} I_s + \int_t^{t-\tau} [\beta_1 I_m(u)S(u) + k_1 I_m(u)v(u) \\ & + \beta_2 I_s(u)S(u) + k_2 I_s(u)V(u)]du, \\ \dot{V} = {} & \left(1 - \frac{S_0}{S}\right)\dot{S} + \left(1 - \frac{V_0}{V}\right)\dot{V} + e^{\mu\tau}\dot{I}_m + e^{\mu\tau}\dot{I}_s + \beta_1 I_m(t)S(t) + k_1 I_m(t)V(t) \\ & + \beta_2 I_s(t)S(t) + k_2 I_s(t)V(t) - \beta_1 I_m(t-\tau)S(t-\tau) - k_1 I_m(t-\tau)V(t \\ & -\tau) - \beta_2 I_s(t-\tau)S(t-\tau) - k_2 I_s(t-\tau)V(t-\tau). \end{aligned}$$

After making some simplifications, we get

$$\begin{aligned} \dot{V} = {} & \mu S_0\left(2 - \tfrac{S_0}{S} - \tfrac{S}{S_0}\right) + r S_0\left(3 - \tfrac{S_0}{S} - \tfrac{V}{V_0} + \tfrac{S}{S_0}\tfrac{V_0}{V}\right) \\ & + I_m\left[\beta_1 S_0 + k_1 V_0 - (\gamma_1 + \beta_3)e^{\mu\tau}\right] + I_s(\beta_2 S_0 + k_2 V_0) \\ & - (\gamma_2 + d)e^{\mu\tau} I_s. \end{aligned}$$

From the arithmetic and geometric of meaning, we have

$$2-\frac{S_0}{S}-\frac{S}{S_0}<0 \text{ and } 3-\frac{S_0}{S}-\frac{V}{V_0}+\frac{S}{S_0}\frac{V_0}{V}<0.$$

$\dot{V} < 0$, whenever $R_0 < 1$. Therefore, the disease-free equilibrium is globally asymptotically stable when $R_0 < 1$.

Theorem 2.2. For the proposed model, mild endemic equilibrium, E_1, is globally asymptotically stable whenever $R_{0,2} < 1$.

Proof. Consider the Lyapunov function

$$V = S_1 g\left(\tfrac{S}{S_1}\right) + V_1 g\left(\tfrac{V}{V_1}\right) + e^{\mu\tau} g\left(\tfrac{I_m}{I_{m,1}}\right) + e^{\mu\tau} I_s + \int_t^{t-\tau}\Big[\beta_1 I_{m,1} S_1 g\left(\tfrac{S(u)I_m(u)}{S_1 I_{m,1}}\right)$$
$$+ k_1 I_{m,1} V_1 g\left(\tfrac{V(u)I_m(u)}{V_1 I_{m,1}}\right) + \beta_2 I_s(u)S(u) + k_2 I_s(u)V(u)\Big]du,$$

Taking the derivative of V, we obtain

$$\begin{aligned}\dot{V} &= \left(1-\tfrac{S_1}{S}\right)\dot{S} + \left(1-\tfrac{V_1}{V}\right)\dot{V} + e^{\mu\tau}\left(1-\tfrac{I_{m,1}}{I_m}\right) \dot{I}_m + e^{\mu\tau}\dot{I}_s\\
&+ \beta_1 S_1 I_{m,1}\Big[\tfrac{S(t)I_m(t)}{S_1 I_{m,1}} - \ln\left(\tfrac{S(t)I_m(t)}{S_1 I_{m,1}}\right) - \tfrac{S(t-\tau)I_m(t-\tau)}{S_1 I_{m,1}}\\
&+ \ln\left(\tfrac{S(t-\tau)I_m(t-\tau)}{S_1 I_{m,1}}\right)\Big]\\
&+ k_1 I_{m,1} V_1\Big[\tfrac{V(t)I_m(t)}{V_1 I_{m,1}} - \ln\left(\tfrac{V(t)I_m(t)}{V_1 I_{m,1}}\right) - \tfrac{V(t-\tau)I_m(t-\tau)}{V_1 I_{m,1}}\\
&+ \ln\left(\tfrac{V(t-\tau)I_m(t-\tau)}{V_1 I_{m,1}}\right)\Big] + \beta_2 I_s(t)S(t) + k_2 I_s(t)V(t)\\
&- \beta_2 I_s(t-\tau)S(t-\tau) - k_2 I_s(t-\tau)V(t-\tau).\end{aligned}$$

Making some simplifications, we get

$$\begin{aligned}\dot{V} \le \mu S_1\left(2-\tfrac{S_1}{S}-\tfrac{S}{S_1}\right) &+ r S_1\left(3-\tfrac{S_1}{S}-\tfrac{V}{V_1}+\tfrac{S}{S_1}\tfrac{V_1}{V}\right)\\
&- \beta_1 I_{m,1} S_1\Big[g\left(\tfrac{S_1}{S(t)}\right) + g\left(\tfrac{I_m(t-\tau)S(t-\tau)}{I_m S_1}\right)\Big]\\
&+ I_s e^{\mu\tau}\left(\tfrac{\lambda e^{-\mu\pi}}{(\gamma_2+d)(r+\mu)}\Big[\beta_2 + \tfrac{k_2 r}{1-k_1-k_2}\Big]-1\right).\end{aligned}$$

From the arithmetic and geometric of meaning, we have

$$2-\frac{S_1}{S}-\frac{S}{S_1}<0 \text{ and } 3-\frac{S_1}{S}-\frac{V}{V_1}+\frac{S}{S_1}\frac{V_1}{V}<0.$$

Hence, $\dot{V} < 0$, whenever $R_{0,2} < 1$. Therefore, the mild endemic equilibrium is globally asymptotically stable when $R_{0,2} < 1$.

Theorem 2.3. For the proposed model, severe endemic equilibrium, E_2, is globally asymptotically stable whenever $R_{0,1} < 1$.

Proof. Consider the Lyapunov function

$$V = S_2 g\left(\frac{S}{S_2}\right) + V_2 g\left(\frac{V}{V_2}\right) + e^{\mu\tau} g\left(\frac{I_s}{I_{s,2}}\right) + e^{\mu\tau} I_m + \int_t^{t-\tau} \Big[\beta_2 I_{s,2} S_2 g\left(\frac{S(u) I_s(u)}{S_2 I_{s,2}}\right)$$
$$+ k_1 I_{m,1} V_1 g\left(\frac{V(u) I_s(u)}{V_2 I_{s,1}}\right) + \beta_1 I_m(u) S(u) + k_1 I_m(u) V(u)\Big] du,$$

Taking the derivative of V, we obtain

$$\dot{V} = \left(1 - \frac{S_2}{S}\right)\dot{S} + \left(1 - \frac{V_2}{V}\right)\dot{V} + e^{\mu\tau}\left(1 - \frac{I_{s,2}}{I_s}\right) I_m + e^{\mu\tau} \dot{I}_s$$
$$+ \beta_2 S_2 I_{s,2}\left[\frac{S(t) I_s(t)}{S_2 I_{s,2}} - \ln\left(\frac{S(t) I_s(t)}{S_2 I_{s,2}}\right) - \frac{S(t-\tau) I_s(t-\tau)}{S_2 I_{s,2}} + \ln\left(\frac{S(t-\tau) I_s(t-\tau)}{S_2 I_{s,2}}\right)\right]$$
$$+ k_2 I_{s,2} V_2\left[\frac{V(t) I_s(t)}{V_2 I_{s,2}} - \ln\left(\frac{V(t) I_s(t)}{V_2 I_{s,2}}\right) - \frac{V(t-\tau) I_s(t-\tau)}{V_2 I_{s,2}} + \ln\left(\frac{V(t-\tau) I_s(t-\tau)}{V_2 I_{s,2}}\right)\right]$$
$$+ \beta_1 I_m(t) S(t) + k_1 I_m(t) V(t) - \beta_1 I_m(t-\tau) S(t-\tau) - k_1 I_m(t-\tau) V(t-\tau).$$

Making some simplifications, we get

$$\dot{V} \leq \mu S_2\left(2 - \frac{S_2}{S} - \frac{S}{S_2}\right) + r S_2\left(3 - \frac{S_2}{S} - \frac{V}{V_2} + \frac{S}{S_2}\frac{V_2}{V}\right)$$
$$- \beta_2 I_{s,2} S_2\left[g\left(\frac{S_2}{S(t)}\right) + g\left(\frac{I_s(t-\tau) S(t-\tau)}{I_s S_2}\right)\right]$$
$$+ I_s e^{\mu\tau}\left(\frac{\lambda e^{-\mu\pi}}{(\gamma_1 + \beta_3)(r+\mu)}\left[\beta_1 + \frac{k_1 r}{1 - k_1 - k_2}\right] - 1\right).$$

From the arithmetic and geometric of meaning, we have

$$2 - \frac{S_2}{S} - \frac{S}{S_2} < 0 \text{ and } 3 - 3 - \frac{S_2}{S} - \frac{V}{V_2} + \frac{S}{S_2}\frac{V_2}{V} < 0.$$

Hence, $\dot{V} < 0$, whenever $R_{0,1} < 1$. Therefore, the severe endemic equilibrium is globally asymptotically stable when $R_{0,1} < 1$.

3 Numerical Simulations

In this section, numerical simulations are made and the following figures are obtained with the help of MatLab to see the effect of vaccine with different proportion of vaccinated individuals.

Figures are constructed by changing recruitment rate, λ, and rate of susceptible individuals that are vaccinated, r, to see their effect on the disease. Recruitment rates are selected as 200 and 300. Proportions of vaccinated individuals are selected as 0.2 and 0.5.

In Fig. 1 and Fig. 2, recruitment rates are changed with constant vaccination, 0.2. This means it is assumed that 20% of susceptible individuals are vaccinated. Also, in Fig. 1 it is assumed that mobility is higher and less people obey the rules (putting mask, social distance, and hygiene) with comparison to Fig. 2. In Fig. 3 and Fig. 4, recruitment rates are changed with constant vaccination, 0.5 which means that 50% of susceptible individuals are vaccinated. In Fig. 3, again, mobility and individuals that

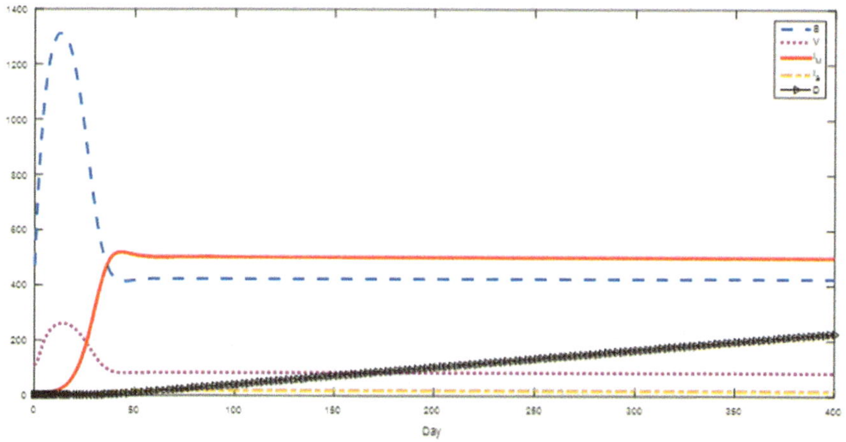

Fig. 1 $\lambda = 300$, $r = 0.2$, $\mu = 0.05$, $\beta_1 = 0.001$, $\beta_2 = 0.00001$, $\beta_3 = 0.01$, $k_1 = 0.00001$, $k_2 = 0.000001$, $\tau = 4.6\,days$, $R_{0,1} = 3.45$, $R_{0,2} = 0.065$

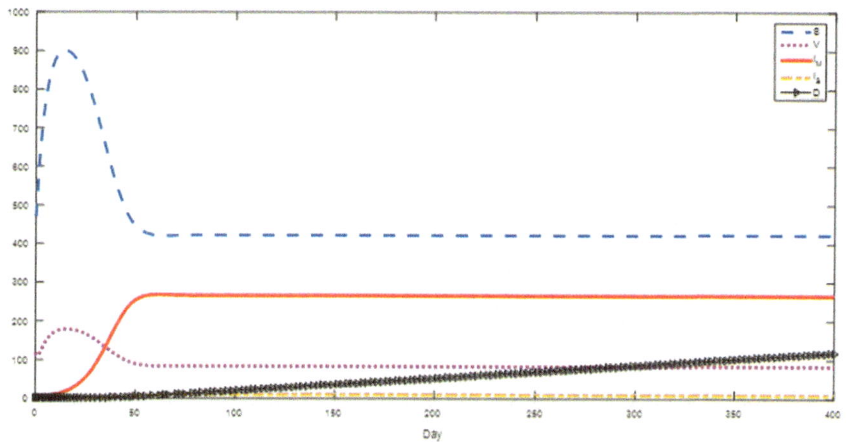

Fig. 2 $\lambda = 200$, $r = 0.2$, $\mu = 0.05$, $\beta_1 = 0.001$, $\beta_2 = 0.00001$, $\beta_3 = 0.01$, $k_1 = 0.00001$, $k_2 = 0.000001$, $\tau = 4.6\,days$, $R_{0,1} = 2.30$, $R_{0,2} = 0.043$

do not obey the rules are higher than Fig. 4. So, with those comparisons, importance of recruitment rates can be seen.

On the other hand, with the comparison of Fig. 1 and Fig. 3, or Fig. 2 and Fig. 4, importance of proportion of vaccinated individuals is noticeable since r is changed in these figures.

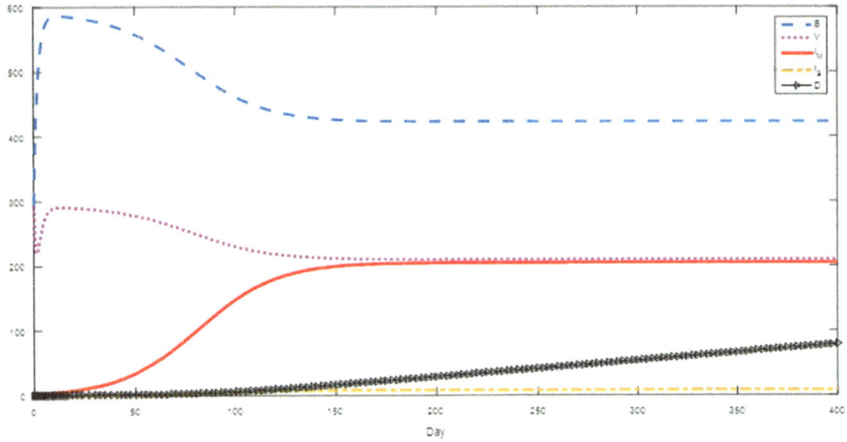

Fig. 3 $\lambda = 300, \quad r = 0.5, \quad \mu = 0.05, \quad \beta_1 = 0.001, \quad \beta_2 = 0.00001, \quad \beta_3 = 0.01, k_1 = 0.00001, \quad k_2 = 0.000001, \quad \tau = 4.6\,days, \quad R_{0,1} = 1.40, \quad R_{0,2} = 0.027$

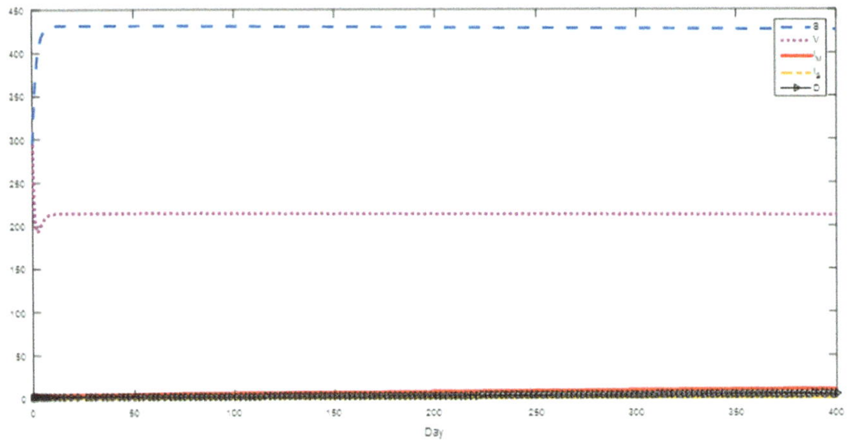

Fig. 4 $\lambda = 200, \quad r = 0.5, \quad \mu = 0.05, \quad \beta_1 = 0.001, \quad \beta_2 = 0.00001, \quad \beta_3 = 0.01, k_1 = 0.00001, \quad k_2 = 0.000001, \quad \tau = 4.6\,days, \quad R_{0,1} = 0.94, \quad R_{0,2} = 0.018$

4 Conclusion

In this paper, a mathematical model is proposed to see the effect of vaccine on Covid-19. Equilibrium points of the system are analyzed and obtained. Afterwards, basic reproduction number, namely R_0, is constructed by using Next Generation Matrix method. Then, necessary theorems and their proofs are given for stability analysis of the model. Finally, numerical simulations of the model are given with the help of MatLab.

When the figures are examined, it is so neat to say that mobility is very important for the control of the spread of Covid-19. For the situation where the movement is restricted, people follow the rules more, and 20% of the population is vaccinated, R_0 value and infected individuals with mild infection decreases. Beside this, approximately 200,000 people can be prevented from contamination (Fig. 1 and Fig. 2). Similar results are obtained from the comparison of Fig. 3 and Fig. 4. When Fig. 1 and Fig. 3 are compared, it can be seen that restrictions and mobility kept constant while proportion of vaccinated individuals are changed. Even with higher vaccinated individuals, namely 50%, R_0 value stays above 1 with less infected individuals. On the other hand, in Fig. 4, mobility is decreased with more people obeying the hygiene rules with 50% vaccination in population and as a result R_0 value decreased below 1, infected individuals decreased and death rates became almost zero.

As a conclusion, for the control of the disease, proportion of vaccination in a population is important but not enough by itself. Mobility and restrictions are as important as proportion of vaccination, even more important. Hence, it can be said that the epidemic can be overcome with adequate vaccination while still following the protection rules.

References

Baba IA, Kaymakamzade B, Hincal E (2018) Two-strain epidemic model with two vaccinations. Chaos Solitons Fractals 106:342–348

Çetin E, Kiremitçi B, Yurt İ (2008) Matematiksel Epidemiyoloji: Pandemik A/H1N1 Gribi Vakası. İstanbul Üniversitesi İşletme Fakültesi Dergisi 38(2):197–209. Turkish

Chadla MS, Potdar VA, Saha S, et al (2015) Dynamics of influenza seasonality at sub-regional levels in India and implications for vaccination timing. PLoS One 10(5):e012412

Dauhoo M, Dumas L, Gabriel P et al (2018) An introduction to the basic reproduction number in mathematical epidemiology. ESAIM Proc Surv 62:123–138

Diekmann O, Heesterbeek J, Metz J (1990) On the definition and the computation of the basic reproduction ratio, in models for infectious diseases in heterogeneous populations. J Math Biol 28:365–382

Encyclopedia Britannica [Internet]. Vaccine; [about 17 screens], Scotland [cited 9 Dec 2020]. https://www.britannica.com/science/vaccine

Kansagra SM, Farley TA (2009) The modern crystal ball: influenza forecasting with mathematical models. Ann Intern Med 151(12):886–887

Kaymakamzade B, Baba IA, Hincal E (2016) Stability analysis of Osaltamivir-resistant influenza virus model. Procedia Comput Sci 102:333–341

Lahariya C (2016) Vaccine epidemiology: a review. J Family Med Prim Care 5(1):7–15

Mathews JD, Chesson JM, McCaw JM et al (2009) Understanding influenza transmission, immunity and pandemic threats. Influenza Other Respir Viruses 3(4):143–149

Musa SS, Zhao S, Chan HS et al (2019) A mathematical model to study the 2014–2015 large-scale dengue epidemics in Kaohsiung and Tainan cities in Taiwan, China. Math Biosci Eng 16(5):3841–3863

Saha S, Chadla M, Almamun A et al (2014) Influenza seasonality and vaccination timing in tropical and subtropical areas of Southern and South-Eastern Asia. Bull World Health Organ 92(5):318–330

Scherer A, McLean A (2002) Mathematical models of vaccination. Br Med Bull 62(1):187–199

Tuite AR, Tien J, Eisenberg M et al (2011) Cholera epidemic in Haiti, 2010: using a transmission model to explain spatial spread of disease and identity optimal control interventions. Ann Intern Med 154(9):593–601
van den Driessche P, Watmough J (2002) Reproduction numbers and sub-threshold endemic equilibria for compartmental models of disease transmission. Math Biosci 180(1–2):29–48
Wang Z, Zhao D, Wang L et al (2015) Immunity of multiplex networks via acquaintance vaccination. Europhys Lett 112(4):48002
WebMD [Internet]. Coronavirus and COVID-19: What You Should Know; [about 38 screens], New York. https://www.webmd.com/lung/coronavirus. Accessed 17 Dec 2020
Worldometer [Internet]. COVID-19 Coronavirus Pandemic; [about 35 screens], United States. https://www.worldometers.info/coronavirus/. Accessed 11 Feb 2021
Zhen W, Chris TB, Samit B et al (2016) Statistical physics of vaccination. Phys Rep 664:1–113

Importance of Carrying Capacity While Fighting with COVID-19

Bilgen Kaymakamzade, Evren Hincal, and Nezihal Gokbulut

Abstract Carrying capacity plays an important role in the research area as it determines the maximum capacity of the system. In epidemiology, vaccine is an efficient control strategy that affects the carrying capacity of health infrastructures. In this study, the sitution of COVID-19 disease in North Cyprus is examined. With the data of country carrying capacity is calculated and compared with different percentage of vaccines to see what should be done for taking the disease under control in the country. The results showed that vaccination should be applied to individuals briskly to reach the herd immunity threshold which should be at least 85% in North Cyprus.

Keywords COVID-19 · Carrying capacity · Herd immunity · Vaccine

1 Introduction

The world has been facing with so many outbreaks since the beginning. When an infectious disease spreads over a specific region, it can be defined as an epidemic while an infectious disease can be referred as a pandemic if it spreads in multiple regions. Plague, Cholera, Influenza, SARS-CoV, MERS-CoV, and SARS-CoV-2 are some of the examples in the history of pandemics in world. According to World Health Organization (WHO), plague has a case-fatality ratio of 30% to 60% with 1–7 days of incubation period. It has caused at least three pandemics in history (Plague 2017). In the past, seven pandemics happened because of the cholera disease. Statistics of WHO showed that each year 1,3–4 million people catch the disease and approximately 21,000–143,000 people died because of cholera (Cholera 2021). Influenza is another pandemic which is the origin of at least 5 pandemics in previous years (Piret and Boivin 2021). Only in 1918, Spanish influenza took 50 million people's lives with approximately 500 million infections (Abdelrahman et al. 2020).

B. Kaymakamzade · E. Hincal · N. Gokbulut (✉)
Department of Mathematics, Near East University, 99138 Nicosia, Cyprus
e-mail: nezihal.gokbulut@neu.edu.tr

DESAM Research Institute, Near East University, 99138 Nicosia, Cyprus

Mathematical Research Center, Near East University, 99138 Nicosia, Cyprus

D. Uzun Ozsahin et al. (eds.), *Decision Analysis Applied to the Field of Environmental Health*, Professional Practice in Earth Sciences,
https://doi.org/10.1007/978-3-030-96682-9_9

SARS-CoV, MERS-CoV, and SARS-CoV-2 are all belong to the coronavirus family. Two of them, SARS-CoV and MERS-CoV, stayed as an epidemic while SARS-CoV-2 became a pandemic. SARS-CoV epidemic started in South China and spread 29 countries with approximately 8000 infections and 774 deaths in November, 2002. In Saudi Arabia an epidemic, known as MERS-CoV, began in September 2012 and 27 countries are affected. Because of MERS-CoV, 2519 people infected and 866 people died (Abdelrahman et al. 2020). In December 2019, a novel coronavirus, SARS-CoV-2, has first appeared in Wuhan, China and spread all around the world in a short time. As of 7th of June, more than 4 million people passed away because of the disease and approximately 186 million people infected (COVID-19 CORONAVIRUS PANDEMIC 2021). These values are increasing day by day and this situation can be concluded as SARS-CoV-2 is much more contagious than the other two coronaviruses (Abdelrahman et al. 2020; House et al. 2021). The disease caused by the virus SARS-CoV-2 is named as COVID-19 by WHO (Chen et al. 2020). At the beginning, COVID-19 was only transmitted via small droplets of infected individual when he/she sneezes, coughs or spokes. However, after March 2020, WHO stated that the disease can also be transmitted through airborne (Transmission of SARS-CoV-2: implications for infection prevention precautions 2020). Its common symptoms are very similar with the symptoms of influenza such as high fever, aches, cough, diarrhea, etc.

Nowadays, countries are experiencing serious economical and health problems because of COVID-19. Almost every country is trying to apply efficient strategies like lockdowns, restrictions, closure of borders, etc. with the purpose of taking the disease under control. Herd immunity and vaccine has been two of the most discussed control strategies for the disease. Herd immunity can be defined as an indirect protection of susceptible individuals from infection via enough number of infected individuals or vaccination. For this, there is a herd immunity threshold that should be reached in population (Randolph and Barreiro 2020). This can be found with the help of mathematical modelling.

Mathematical modelling is a very decisive way that can be used for controlling the spread of infectious diseases. Basically, modelling divides the defined population into sensible number of compartments, analyzes the relationship of compartments and aims to determine control strategies for overcoming the disease. Constructing the basic reproduction number, denoted by R_0, of the disease is an efficient way while controlling infectious diseases. R_0 can be defined as the number of secondary infections per primary infection. If R_0 values are less than 1, then it means that the pursued policy is enough and efficient for the control of the disease. Otherwise, it means much more applications are needed for controlling the disease (Guerra et al. 2017). Beside this, carrying capacity is an important measure in infectious diseases for the health infrastructure. In epidemiology, carrying capacity can be defined as the absolute maximum number of infected individuals in a population based on the amount of the limited health sources (Hale and McCarthy).

Mathematical modelling is very popular in infectious diseases; (Towers and Feng 2009) and (Matrajt et al. 2021) are some models that have studied vaccine while the dynamics of infectious diseases are studied in Mathews et al. (2009) and (Kaymakamzade et al. 2016). Hincal et al. (2020) is another study that studied the herd immunity idea. Carrying capacity is concentrated in the models (Duarte et al. 2003) and (DeAngelis et al. 2020).

With the first COVID-19 case seen on 10th of March in North Cyprus, the government applied many different strategies including lockdowns and closure of airports and borders. These rules are followed by looser restrictions such as less mobility, online education, regional lockdown, quarantine of contacted individuals etc. Many mathematical models are constructed for the control of the spread of COVID-19 in North Cyprus according to the existing situation in the island (Hincal et al. 2020; Hincal et al. 2021). Herd immunity was studied in North Cyprus and the results showed that herd immunity did not hold in country through infection (Hincal et al. 2020).

This study captures the dates between February - July 2021 to analyze the situation under those conditions (the existing situation). Population in North Cyprus is taken as 382,000. Between the given dates, applied restrictions include regional lockdown, lockdown of island, limited number of passengers, changing mobility with percentages, and changing vaccine proportions.

This study aims to show the importance of herd immunity through vaccination with the help of carrying capacity. According to the Robert Koch Institute, the people in the range of 18–59 as an age should be vaccinated with 85% while the people over 60 ages should get the vaccine at least 90% for the herd immunity through vaccination. In our country, because of continuous internal-external migrations, the population can not be identified exactly which makes hard to decide the percentage of herd immunity via vaccination in North Cyprus. In the study, carrying capacity of health infrastructures is identified with different percentages of vaccination.

2 Model Formulation

A new mathematical model is constructed by taking into consideration of struggle of Covid-19 in North Cyprus Whole population, denoted by N, is divided into 7 compartments, according to the progress of the disease as susceptible individuals, denoted by S, vaccinated individuals, denoted by V, quarantined individuals, denoted by Q, mild infected individuals, denoted by I_M, severe infected individuals, denoted by I_S, recovered individuals, denoted by R, and deaths because of Covid-19, denoted by D. Thus, $N(t) = S(t) + V(t) + Q(t) + I_M(t) + I_S(t) + R(t) + D(t)$, at time t.

In the study it is assumed that only severe infected individuals can die because of the disease. Beside this, it is assumed that severe infection can occur from mild infection only. So, there is no direct transition from the compartment S to the compartment I_S.

The model is constructed with Ordinary Differential Equations (ODEs) as follows:

$$\frac{dS}{dt} = \Lambda - (\beta_M I_M + \beta_S I_S)S + (1-\theta)Q - (\eta + p)S,$$

Table 1 Descriptions of parameters used in the model

Parameters	Descriptions
Λ	Recruitment rate
β_M	Mild infection rate for susceptible individuals
β_S	Severe infection rate for susceptible individuals
θ	Positivity rate of quarantined individuals
η	Natural death rate
p	Vaccination rate of susceptible individuals
k_M	Mild infection rate for vaccinated individuals
k_S	Severe infection rate for vaccinated individuals
σ	Rate of (transferring) mild infected individuals to be severe infected
ε_M	Recovery rate of mild infected individuals
ε_S	Recovery rate of severe infected individuals
d	Death rate of infected individuals

$$\begin{aligned}
\frac{dV}{dt} &= pS - (k_M I_M + k_S I_S)V - \eta V, \\
\frac{dQ}{dt} &= (1-c)(\beta_M I_M + \beta_S I_S)S - Q, \\
\frac{dI_M}{dt} &= c(\beta_M I_M + \beta_S I_S)S + \theta Q - (\sigma + \varepsilon_M)I_M + (k_M I_M + k_S I_S)V, \\
\frac{dI_S}{dt} &= \sigma I_M - \varepsilon_S I_S - dI_S, \\
\frac{dR}{dt} &= \varepsilon_M I_M + \varepsilon_S I_S, \\
\frac{dD}{dt} &= dI_M.
\end{aligned}$$

The descriptions of parameters are given in Table 1.

3 Model Analysis

Theorem 3.1. Let $(S, V, Q, I_M, I_S, R, D)$ be the solution of the given system with initial conditions $S \geq 0$, $V \geq 0$, $Q \geq 0$, $I_M \geq 0$, $I_S \geq 0$, $R \geq 0$, $D \geq 0$. Then, the set.

$$\Pi = \left\{(S, V, Q, I_M, I_S, R, D) \in R^7_+ : S + V + Q + I_M + I_S + R + D \leq \Lambda\right\}$$

is positive, invariant, and all of the solutions in R^7_+ stay in Π with respect to the given system.

Proof. As a result of adding all of the terms on the right side of the system,

$$\frac{dN}{dt} = \Lambda - \eta(S + V) - d(I_M + I_S)$$

is obtained. From this, it is obvious that $\frac{dN}{dt} \leq \Lambda$. Integrating both sides, $Ne^t \leq \Lambda e^t + b$ is get for some arbitrary constant b. By using Rota and Birkhoff for the differential inequality, it can be found that $0 \leq N \leq \Lambda$ holds as t tends to ∞. Hence, N approaches Λ as t tends to infinity and so the solutions of the given system enters the region Π. It is guaranteed that the model is biologically feasible and it is enough to consider the dynamics on the model in Π.

3.1 The Disease Free Equilibrium Point (DFE)

The constructed model reveals a unique disease-free equilibrium point (DFE) which is

$$C^0 = \left(S_0, V_0, Q_0, I_{M,0}, I_{S,0}, R_0, D_0\right) = \left(\frac{\Lambda}{\eta + p}, \frac{p\Lambda}{\eta(\eta + p)}, 0, 0, 0, 0, 0\right).$$

It is clear that C^0 attracts the region so that

$$C^0 = \left\{\left(S_0, V_0, Q_0, I_{M,0}, I_{S,0}, R_0, D_0\right) \in R_7^+ : Q_0 = I_{M,0} = I_{S,0} = R_0 = D_0 = 0\right\}.$$

3.2 The Basic Reproduction Number

The basic reproduction number of an infectious disease can be simply defined as the average number of secondary infections occurred by a primary infection with the assumption that all population is susceptible, and it is denoted by R_0. In order to say that any disease is under control, the value of basic reproduction number should be less than 1.

The basic reproduction number of Covid-19 for the presented model is calculated by using the next generation matrix method. Firstly, the following matrices are constructed by using the system of ODE's:

$$f = \begin{bmatrix} c(\beta_M I_M + \beta_S I_S)S + \theta Q + (k_M I_M + k_S I_S)V \\ 0 \end{bmatrix},$$

$$v = \begin{bmatrix} (\sigma + \varepsilon_M)I_M \\ \varepsilon_S I_S + dI_S - \sigma I_M \end{bmatrix}.$$

$$+ -(1-\theta)Q\frac{S_0}{S} + I_M(\beta_M S_0 + k_M V_0 - \varepsilon_M)$$
$$+ I_S(\beta_S S_0 + k_S V_0 - \varepsilon_S - d).$$
$$\dot{V} = S_0\eta\left(2 - \frac{S_0}{S} - \frac{S}{S_0}\right) + pS_0\left(3 - \frac{S_0}{S} - \frac{V}{V_0} - \frac{V_0}{V}\frac{S}{S_0}\right)$$
$$+ -(1-\theta)Q\frac{S_0}{S} + I_M\varepsilon_M\left(\frac{\beta_M S_0 + k_M V_0}{\varepsilon_M} - 1\right)$$
$$+ I_S(\varepsilon_S + d)\left(\frac{\beta_S S_0 + k_S V_0}{\varepsilon_S + d} - 1\right)$$
$$\leq S_0\eta\left(2 - \frac{S_0}{S} - \frac{S}{S_0}\right) + pS_0\left(3 - \frac{S_0}{S} - \frac{V}{V_0} - \frac{V_0}{V}\frac{S}{S_0}\right) + -(1-\theta)Q\frac{S_0}{S}$$
$$+ I_M\varepsilon_M\left(\frac{c\beta_M S_0 + k_M V_0}{\varepsilon_M + \sigma} - 1\right)$$
$$+ I_S(\varepsilon_S + d)\left(\frac{\beta_S S_0 + k_S V_0}{\varepsilon_S + d} - 1\right).$$

From arithmetic and geometric inequality, we have $2 - \frac{S_0}{S} - \frac{S}{S_0} \leq 0$ and $3 - \frac{S_0}{S} - \frac{V}{V_0} - \frac{V_0}{V}\frac{S}{S_0} \leq 0$. Therefore, when $R_0 < 1$, $\frac{c\beta_M S_0 + k_M V_0}{\varepsilon_M + \sigma} + \frac{\beta_S S_0 + k_S V_0}{\varepsilon_S + d} < 1$ which means that $\dot{V} < 0$. Hence, the DFE E_0 is globally asymptotically stable if $R_0 < 1$.

4 Numerical Simulations

In this section, numerical simulations are obtained by using real parameters and the model with MatLab.

Figure 1 represents the weekly cases in North Cyprus per 100,000. According to the World Health Organization (WHO), the value of cases per 100,000 is evaluated as follows:

- $\leq 20 \Rightarrow$ low risk,
- Between 21 and 50 $\Rightarrow$ moderate risk,
- Between 51 and 100 $\Rightarrow$ high risk,
- $\geq 100 \Rightarrow$ ultrahigh risk.

Due to the evaluation of WHO, North Cyprus is in the ultrahigh risk area.

Figure shows the comparison of cases and results of the model. In the figure, blue line represents the occurred cases on those days while the dashed red line represents expected cases with the model. The difference between the lines can be explained by the increase in humidity at the beginning of Summer which may affect the results of the model because of the negative effect of humidity on the spread of Covid-19 (Hincal et al. 2021).

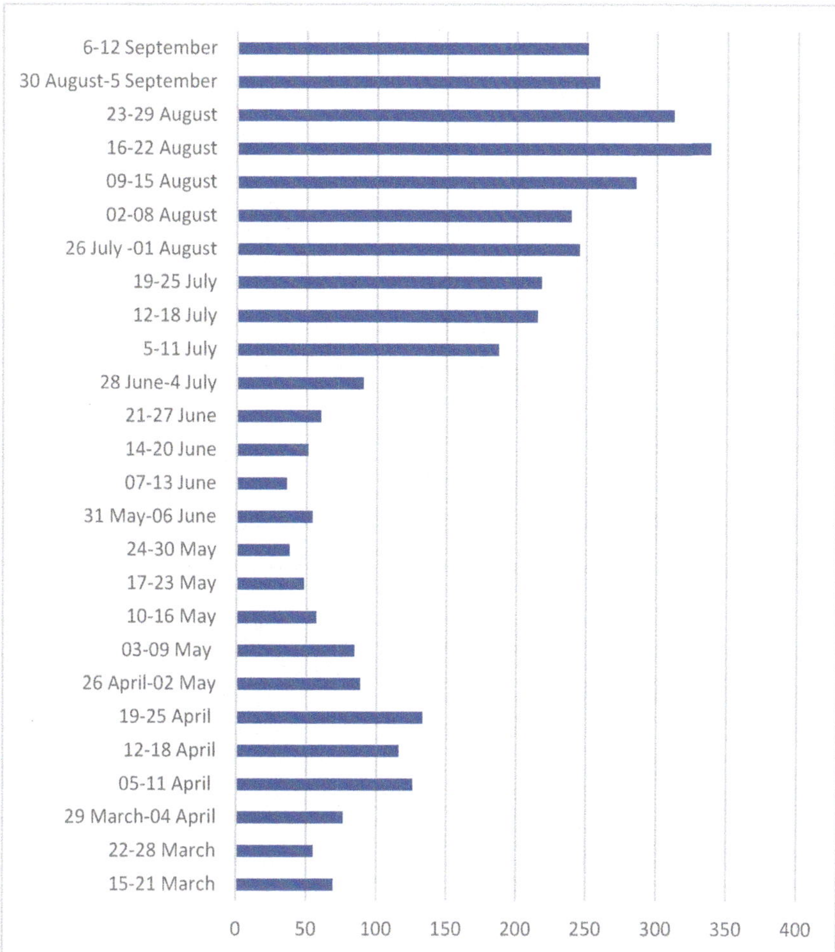

Fig. 1 Weekly cases per 100,000

Figure 3 represents the pattern of active cases in North Cyprus. Its peak value is 2284 which is the carrying capacity threshold.

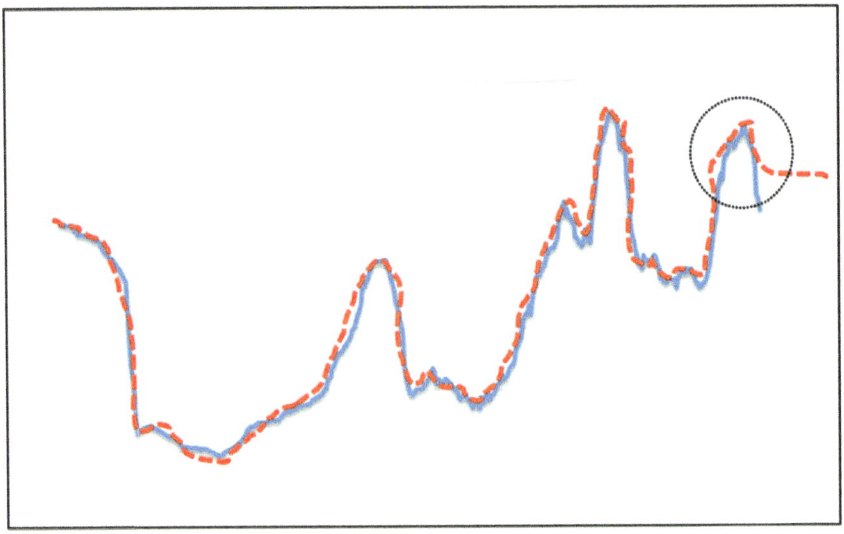

Fig. 2 Comparison of active cases and results of the model between 11 March 2020 - 4 May 2021

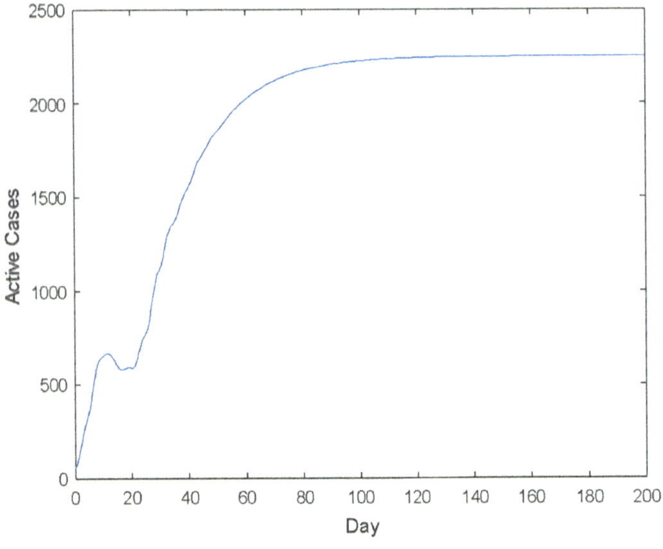

Fig. 3 The pattern of active cases in North Cyprus (Tarih ver) and carrying capacity (maximum 2284 cases)

5 Conclusion and Discussion

In this paper, the constructed model has showed that the spread of Covid-19 in North Cyprus is not under control. From the pattern of cases per 100,000, given in Fig. 1, it can be easily seen that it had been dropped for a short time due to restrictions. However, with lifted restrictions the value of cases per 100,000 has started to increase with a peak value 338, under the current circumstances. In May 2021, the model is presented to the government to show what is expected to happen. The figures obtained from the model displayed that with no restrictions R_0 value will exceed 8.0 and the carrying capacity, which is approximately 2284, will be exceeded. Simulations are given in Fig. 2 and Fig. 3. However, the results of the model are neglected because of the drop of positive cases during those days (beginning of Summer 2021). The results of the model and the prediction of cases (as a result of the drop) are shown in Fig. 2. This drop can be explained by the increase in humidity at the end of May since it is proved that humidity has a negative effect on the spread of Covid-19. In other words, it reduces the spread of Covid-19. On the other hand, this drop (in positive cases) has lasted in a short time because of continuous change in variants of Covid-19 in North Cyprus.

Increase in R_0 values cause a serious rise in active cases and in a short time period, these may lead carrying capacity to be surpassed if no restrictions would not apply in the country. Under these circumstances or without any precautions, carrying capacity is expected to be exceeded which may damage the healthcare system and people, and the pandemic will be much more unrestrainable. Beside this, under the current conditions, herd immunity is not possible to reach.

References

Abdelrahman Z, Li M, Wang X (2020) Comparative review of SARS-CoV-2, SARS-CoV, MERS-CoV, and influenza a respiratory viruses. Front Microbiol

Chen N et al (2020) Epidemiological and clinical characteristics of 99 cases of 2019 novel coronavirus pneumonia in Wuhan, China: a descriptive study. Lancet 395(10223):507–513

Cholera (5 February 2021). World Health Organization. https://www.who.int/news-room/fact-sheets/detail/cholera

COVID-19 CORONAVIRUS PANDEMIC (8 July 2021) Worldometer. https://www.worldometers.info/coronavirus/

DeAngelis D, Zhang B, Ni W-M, Wang Y (2020) Carrying capacity of a population diffusing in a heterogeneous environment. Mathematics 8

Duarte P, Meneses R, Hawkins AJ, Zhu M, Fang J, Grant J (2003) Mathematical modelling to assess the carrying capacity for multi-species culture within coastal waters. Ecol Model 168(1–2):109–143

Guerra FM et al (1 December 2017) The basic reproduction number (R0) of measles: a systematic review. Lancet Infect Dis. 17(12):E420–E428. https://doi.org/10.1016/S1473-3099(17)30307-9

Hale MB, McCarthy LM (n.d.) An introduction to population ecology - the logistic growth equation. MAA Publications

Hincal E, Kaymakamzade B, Gokbulut N (2020) Basic reproduction number and effective reproduction. Bull Karaganda Univ 86–95. https://doi.org/10.31489/2020M3/86-95
Hincal E, Kaymakamzade B, Gokbulut N (2021) Humidity level on COVID-19 with control strategies. Int J Appl Math 34:795–802
House NN, Palissery S, Sebastian H (2021) Corona viruses: a review on SARS, MERS and COVID-19. Microbiol Insights 14
Kaymakamzade B, Baba IA, Hınçal E (2016) Global stability snalysis of Oseltamivir–resistant influenza virus model. Procedia Comput Sci 102:333–341. https://doi.org/10.1016/j.procs.2016.09.409
Mathews JD, Chesson JM, McCaw JM, McVernon J (2009) Understanding influenza transmission, immunity and pandemic threats. Influenza Respir Viruses 3(4):143–149
Matrajt L, Eaton J, Leung T, Brown ER (2021) Vaccine optimization for COVID-19: who to vaccinate first? Sci Adv 7(6)
Piret J, Boivin G (2021) Pandemics throughout history. Front Microbiol
Plague (31 October 2017) World Health Organization. https://www.who.int/news-room/fact-sheets/detail/plague
Randolph HE, Barreiro LB (2020) Herd Immunity: understanding COVID-19. Immunity 52(5):737–741. https://doi.org/10.1016/j.immuni.2020.04.012
Towers S, Feng Z (2009) Pandemic H1N1 influenza: predicting the course of a pandemic and assessing the efficacy of the planned vaccination programme in the United States. Eurosurveillance 14(41):6–8
Transmission of SARS-CoV-2: implications for infection prevention precautions (9 July 2020) World Health Organization. https://www.who.int/news-room/commentaries/detail/transmission-of-sars-cov-2-implications-for-infection-prevention-precautions

Determination of the Epidemic Character of HIV Infections in Children in Turkey Using a Mathematical Model

Nazife Sultanoglu, Farouk Tijjani Saad, Tamer Sanlidag, Bilgen Kaymakamzade, Evren Hincal, and Murat Sayan

Abstract HIV not only affects adults but also infants and children. In 2020, 1.7 million children under the age of 15 were living with HIV and 100,000 HIV-related deaths occurred in children. This study aims to examine the characteristics of the epidemic of HIV-infected children in Turkey by using a mathematical model. To achieve this SIR (Susceptible, Infected, and Removed) model was used which consisted of two equilibrium points: presence or absence of epidemic. The basic reproduction number (R_0) is the number of individuals infected by one person and considered as threshold meaning that if $R_0 < 1$ no epidemic and if $R_0 \geq 1$ an epidemic will occur. In this study, data of 926 HIV-positive children of 0–19 ages who were diagnosed in Turkey from 1985 to 2020 were used. Based on our calculations the size of the HIV epidemic among children in Turkey was calculated to be $R_0 = 1.27$. This value indicated that in Turkey, HIV infection in children currently has epidemic characteristics. In addition, a 50-years perspective simulation was performed using the designed mathematical model. This simulation showed that in 2024, the number of HIV infections in children will start to rise logarithmically. It can be concluded that mathematical modelling can be used to study HIV epidemic characteristics. Based on the results of this study, if the necessary prevention strategies are not followed, it is predicted that a logarithmic rise will occur in the number of HIV-infected children in Turkey in the future.

Keywords HIV · Children · Turkey · Epidemic · Mathematical computing

N. Sultanoglu (✉)
Faculty of Medicine, Department of Medical Microbiology and Clinical Microbiology, Near East University, Nicosia, Cyprus
e-mail: nazife.sultanoglu@neu.edu.tr

N. Sultanoglu · F. T. Saad · T. Sanlidag · B. Kaymakamzade · E. Hincal · M. Sayan
DESAM Research Institute, Near East University, Nicosia, Cyprus

F. T. Saad · B. Kaymakamzade · E. Hincal
Department of Mathematics, Near East University, Nicosia, Cyprus

M. Sayan
Faculty of Medicine, Clinical Laboratory, PCR Unit, Kocaeli University, Kocaeli, Turkey

D. Uzun Ozsahin et al. (eds.), *Decision Analysis Applied to the Field of Environmental Health*, Professional Practice in Earth Sciences,
https://doi.org/10.1007/978-3-030-96682-9_10

1 Introduction

Human Immunodeficiency Virus (HIV) infection continues to be a public health problem that not only affects adults but also infants and children. HIV infects and destroys immune cells leading to progressive deterioration of the immune system. Acquired immunodeficiency syndrome (AIDS) is the most advanced stage of HIV infection and it is defined by the development of opportunistic infections and certain cancers. The progression of HIV infection to AIDS may take 2 to 15 years if not treated; however, the progression to AIDS is much faster in children (WHO 2017).

At present, there is no cure for HIV infection. However, HIV is a manageable chronic disease by using antiretroviral therapy (ART) (Mahungu et al. 2009). ART is a combination of anti-HIV drugs, referred to as the ART regimen. The regimens are patient -specific administered by health practitioners. ART enables HIV—infected people to live longer and healthier lives as well as preventing the transmission of the disease to others by controlling the viral load (Blokzijl 1988).

Also, with proper management, when ART is used regularly during the pregnancy of HIV—infected mothers, transmission from mother to baby transmission is diminished drastically. However, when ART is not administered, there is a high risk that the foetus will acquire HIV from the mother and will have the infection throughout his/her life since the HIV infection is lifelong. Thus, inhibiting the transmission of HIV infection to infants and young children has paramount importance (Irshad et al. 2021).

Once HIV is transmitted to infants and children, is a big threat to their survival. The impact of HIV infection in infants and young children living in urban areas is particularly in Africa is striking. There are many clinical manifestations in HIV infection children. These include eczema, pulmonary interstitial pneumonitis, oral or perineal thrush, recurrent bacterial infections, thrombocytopenia, failure to thrive, and hepatosplenomegaly. Also frequently seen opportunistic infections include Epstein-Barr virus, *Pneumocystis jirovecii* pneumonia, Cryptosporidium diarrhoea and cytomegalovirus.

However, these are frequently seen diseases in children in Africa due to malnutrition and recurrent parasitic infections. Thus, this renders, recognition of the HIV infection in children very difficult in this particular area (Blokzijl 1988).

According to 2020 data published by the World Health Organization (WHO), 37.6 million people are living with HIV around the world. Of these 37.6 million, 1.7 million are children under the age of 15 and approximately 100,000 HIV-related deaths occurred in children in 2020. The transmission of HIV in children may occur in utero, during birth, or while being breastfed, which are referred to as mother to child transmission (also known as vertical transmission). The incidence of HIV infection in children differs from region to region with Sub-Saharan Africa being more affected (WHO 2006, 2021).

Vertical transmission majorly occurs in utero or children at a young age due to breastfeeding. In children aged 10–19 years who are considered adolescents, the transmission trend changes from vertical to horizontal transmission, which includes needle sharing and sexual contact (WHO 2013). In the adolescent phase, the sexual maturation process takes place and they explore relationships, gender norms, and sexuality (Naswa and Marfatia 2010; AVERT 2019).

Transmission through blood products is another way of HIV transmission to occur. However, due to the increased use of reliable methods of screening blood donors for HIV infection, the transmission of HIV via blood products has decreased drastically. Despite the screening procedures, if these tests are not available the risk of transmission through blood will increase the incidence of HIV infection (Blokzijl 1988).

Turkey is one of the countries where HIV infections continue to rise with an increasing trend. According to the Turkish Ministry of Health, from 1985 to 2020, the cumulative total of HIV—positive individuals diagnosed in Turkey was reported to be 26,447. Amongst this number, 926 were recorded to be HIV-positive children aged between 0–19 years. As the age of the children increases, the trend of HIV infection increases. This trend is as follows: there were 102 cases involving children under 1, 68 between 1 and 4 years, 34 between 5 and 9 years, 32 between 10 and 14 years, and 690 cases in children aged between 15 and 19 (Table 1). The reason behind this could be either childhood infection that occurs in the early ages of life but is only diagnosed at older ages or due to children exploring relationships and sexuality. Further study and data should be gathered to state this with greater certainty (Turkey Ministry of Health 2021).

There has been no study conducted previously to study the epidemic character of HIV infections among children in Turkey. This study aims to determine the characteristics of the HIV epidemic in children in Turkey using real cumulative data from 1985 to 2019 on children aged between 0–19 years obtained from the Turkish Ministry of Health by designing a mathematical model. This mathematical model was designed in cooperation with mathematicians, infectious disease experts, paediatricians, and microbiologists to analyse the epidemic of HIV in children in Turkey.

Table 1 Distribution of HIV cases by age groups amongst children and gender

Age group	Male	Female	Total
0	67	35	102
1–4	37	31	68
5–9	22	12	34
10–14	19	13	32
15–19	552	138	690
Cumulative total	962		

(Turkey Ministry of Health 2021).

2 Methods

In this study, the SIR model defined as the S (t) susceptible children, I (t) HIV infected children, and R (t) removed children were used to investigating the epidemic of HIV infected children in Turkey. Susceptible children are not yet infected with HIV but are at risk of acquiring the infection at any time. On the other hand, removed children stand for children who have died naturally (any death that is not due to HIV infection) or due to HIV-related death. These are deducted (removed) from the mathematical model. In this study, the cumulative total number of 926 children diagnosed in Turkey in the years between 1986 and 2020 was used (Turkey Ministry of Health 2021).

The mathematical model used in the current study was adopted from (Saad *et al.* 2019). It consists of three nonlinear ordinary differential equations, which are represented below.

$$\frac{dS}{dt} = \Lambda - \frac{\alpha SH}{N} - \mu S$$

$$\frac{dH}{dt} = \frac{\alpha SH}{N} - (v + \mu)H$$

$$\frac{dR}{dt} = (v + \mu)H + \mu s,$$

$$S(t) > 0, \quad H(t) \geq 0, \quad and \quad R(t) \geq 0.$$

Two equilibrium points which are disease free and endemic equilibrium were found. The representations of the two equilibrium points are shown below:

$$E_0 = \left(S_0 = \frac{\Lambda}{\mu}, H_0 = 0\right),$$

and

$$E_1 = \left(S_1 = \frac{v + \mu}{\alpha}, H_1 = \frac{\Lambda\alpha - \mu v - \mu^2}{\alpha(v + \mu)}\right).$$

The next-generation matrix method was applied for the representation of the basic reproduction number as follows:

$$R_0 = \frac{\alpha\Lambda}{(v + \mu)\mu}$$

The basic reproduction number R_0 is the number of individuals infected by a single infective person in a completely susceptible population. The demonstration of the meaning of R_0 can be visualized in Fig. 1. It is considered as a threshold, meaning

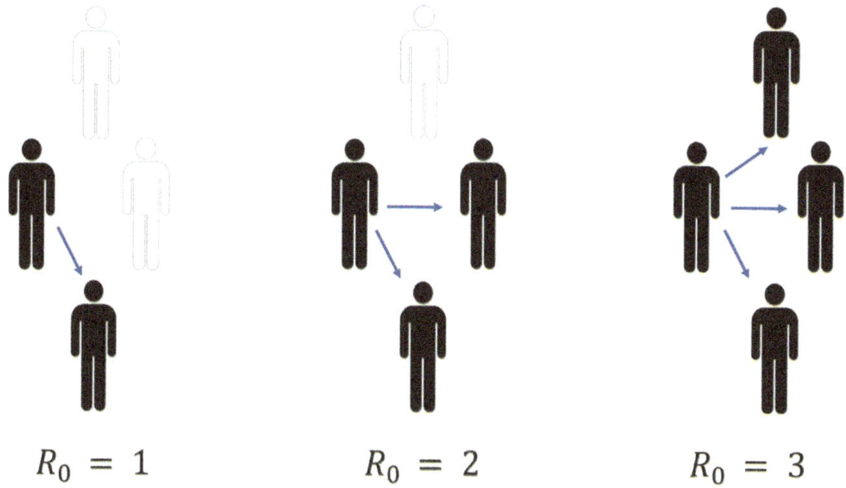

Fig. 1 Basic reproduction number (R_0) demonstration (The figure has been constructed using PowerPoint to demonstrate the R_0)

that if $R_0 \geq 1$ an epidemic will occur, and if $R_0 < 1$ no epidemic will occur.

The stability analysis of the equilibrium points is given with the below theorems.

Theorem 1. The disease-free equilibrium is globally asymptotically stable when $R_0 < 1$.

Theorem 2. The endemic equilibrium E_1 is globally asymptotically stable when $R_0 \geq 1$.

Proof of the theorems was obtained by the Lyapunov function (Saad *et al.* 2019).

The list of variables and parameters that were integrated into the mathematical model for the calculation of the basic reproduction number was obtained from the Turkish Ministry of Health and the Turkish Institute of Health Directorate, as shown in Table 2. After obtaining the value of t R_0, MatLab 2018b software was used to simulate the mathematical model to predict the future trend of HIV in children in Turkey for the next 50 years.

List of variables and parameters used in the mathematical model where some parameters were not directly given by the references used; these were calculated indirectly by mathematicians (TUSEB 2019; Turkey Ministry of Health 2021).

3 Results

The parameters and variables shown in Table 1 were integrated into the basic reproduction formula. From this formula, basic reproduction was calculated to be 1.27.

Table 2 List of variables and parameters used in the mathematical model

Variable/Parameter	Definition	Value
N	Total population of children	22 883 288
S	Number of susceptible children	22 882 612
I	Number of HIV-infected children	926
Λ	Birth rate	1.99
μ	The natural death rate of children	0.20860
A	Infection rate	0.02954
V	The death rate due to HIV infection	0.01400

As the calculated R_0 is greater than 1, this indicates that there is currently an HIV epidemic in Turkey among children aged from 0 to 19 years old.

Furthermore, by using the MatLab 2018b software, the mathematical model was simulated to predict what is likely to happen in the future in terms of the trend of HIV-infected children in Turkey. The obtained results indicate that there will be a logarithmic increase in the next 50 years in HIV-infected children in Turkey, as represented in Fig. 2.

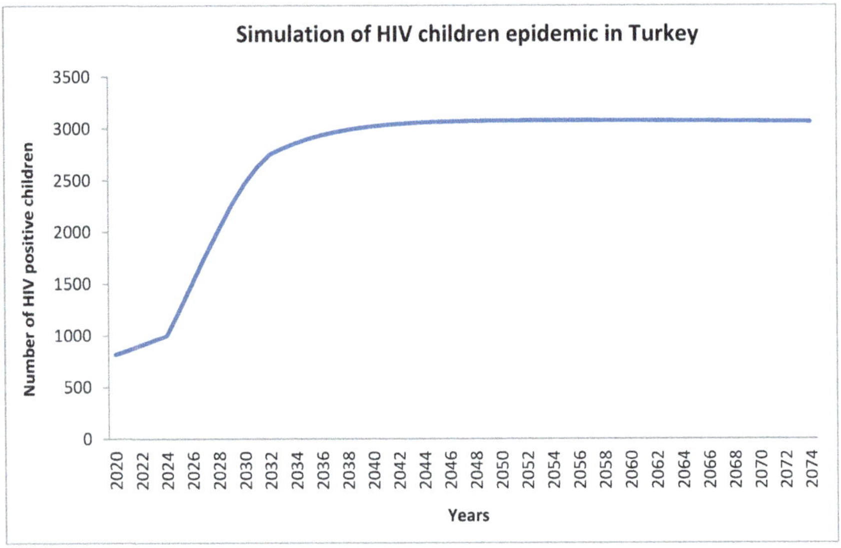

Fig. 2 Simulation showing the predicted trend of HIV infections in children in Turkey for the next 50 years

The mathematical model and the data presented in Table 1 were entered into MatLab 2018b software to predict the trend of HIV infections in children. The results indicated that, if necessary, preventive measures are not taken in Turkey, a rise in the number of HIV-infected children will start in 2024 and will continue to increase logarithmically.

4 Discussion

The designed mathematical model was used to study the epidemic characteristics of HIV infections in children in Turkey. The mathematical model consisted of two equilibrium points: disease-free and endemic. The magnitude of the basic reproduction number determined the stability of the equilibrium points (where equilibrium point stability was controlled by the Lyapunov function). When the value of R_0 is less than 1, this indicates that an epidemic is not likely to occur, whereas if the value of R_0 is greater than 1, this implies that an epidemic is likely to occur.

All the parameters and variables that were integrated into the mathematical model were obtained from the Turkish Ministry of Health and the Turkish Institute of Health Directorate. A total number of 926 HIV-infected children aged between 0 and 19 who were diagnosed between 1985 and 2020 were used to study the epidemic dynamic in children in Turkey. By using the mathematical model and necessary parameters/variables (Table 2), R_0 was calculated to be 1.27. Thus, this indicated that HIV infections in children in Turkey currently have epidemic characteristics. Moreover, a simulation was performed using the designed mathematical model to predict the trends of HIV infections among children in Turkey for the following years. Based on the simulation, it was observed that the number of children infected with HIV will start to rise in 2024 and will continue to increase logarithmically in the following 50 years.

The research team has previously used mathematical modelling to study: the role of awareness in controlling HIV (Kaymakamzade *et al.* 2018), the epidemic dynamics according to each year from 1985 to 2016, and epidemics in Turkey using both adults and children (Sayan *et al.* 2018). Also, others have successfully used mathematical modelling to analyse the HIV epidemic (Naresh et al. 2009; Bozkurt and Peker 2014). It can be concluded that mathematical modelling is useful in analysing the characteristics of HIV infections and can also be used to predict what is likely to happen in the future. The results of this study indicated that HIV infections among children in Turkey have epidemic characteristics. In addition, based on a simulation of the obtained results using the designed mathematical formula, it can be predicted that if the necessary preventive policies are not implemented in Turkey to plan for the next years, the HIV epidemic in children will continue to rise.

Declarations The authors report no conflict of interest and no funding was received from elsewhere to conduct this study. There is no third party work present in this study thus there is no need to obtain third party rights.

References

AVERT (2019) Young people, HIV and AIDS. https://www.avert.org/professionals/hiv-social-issues/key-affected-populations/young-people. Accessed 28 Jan 2020

Blokzijl ML (1988) Human immunodeficiency virus infection in childhood. Ann Trop Paediatr 8(1):1–17. https://doi.org/10.1080/02724936.1988.11748530

Bozkurt F, Peker F (2014) Mathematical modelling of HIV epidemic and stability analysis. Adv Difference Equ 1:95. https://doi.org/10.1186/1687-1847-2014-95

Irshad U, Mahdy H, Tonismae T (2021) HIV in pregnancy. https://www.ncbi.nlm.nih.gov/books/NBK558972/. Accessed 1 July 2021

Kaymakamzade B et al (2018) Role of awareness in controlling HIV/AIDS: a mathematical model. Qual Quant 52:625–637. https://doi.org/10.1007/s11135-017-0640-2

Mahungu TW, Rodger AJ, Johnson MA (2009) HIV as a chronic disease. Clin Med J R Coll Phys Lond 9(2):125–128. https://doi.org/10.7861/clinmedicine.9-2-125

Naresh R, Tripathi A, Sharma D (2009) Modelling and analysis of the spread of AIDS epidemic with immigration of HIV infectives. Math Comput Model 49(5–6):880–892. https://doi.org/10.1016/j.mcm.2008.09.013

Naswa S, Marfatia YS (2010) Adolescent HIV/AIDS: issues and challenges. Indian J Sex Transm Dis AIDS 31(1):1. https://doi.org/10.4103/2589-0557.68993

Saad FT et al (2019) Global stability analysis of HIV+ model, pp 830–839. https://doi.org/10.1007/978-3-030-04164-9_109

Sayan M et al (2018) Dynamics of HIV/AIDS in Turkey from 1985 to 2016. Qual Quant 52:711–723. https://doi.org/10.1007/s11135-017-0648-7

Turkey Ministry of Health (2021) HIV-AIDS Statistics Turkey. https://hsgm.saglik.gov.tr/tr/bulasici-hastaliklar/hiv-aids/hiv-aids-liste/hiv-aids-istatislik.html. Accessed 1 July 2021

TUSEB (2019) Türkiye Sağlık Enstitüleri Başkanlığı. https://www.tuseb.gov.tr/. Accessed 28 Jan 2020

WHO (2006) HIV/AIDS Programme WHAT WE KNOW. https://www.who.int/hiv/toronto2006/Children2_eng.pdf. Accessed 28 Jan 2020

WHO (2013) HIV and Adolescents: Guidance for HIV Testing and Counselling and Care for Adolescents Living with HIV - NCBI Bookshelf. https://www.ncbi.nlm.nih.gov/books/NBK217964/. Accessed 6 July 2021

WHO (2017) HIV/AIDS. https://www.who.int/features/qa/71/en/. Accessed 28 Jan 2020

WHO (2021) Global HIV Programme. https://www.who.int/teams/global-hiv-hepatitis-and-stis-programmes/hiv/strategic-information/hiv-data-and-statistics. Accessed 1 July 2021

Analysis and Simulation of HIV Infected Children Transmission Dynamics in Turkey Using a Mathematical Model

Bilgen Kaymakamzade, Tamer Sanlidag, Nazife Sultanoglu, Farouk Tijjani Sa'ad, Murat Sayan, and Evren Hincal

Abstract Child HIV infections are an important part of HIV prevalence today. The path of infection can occur either vertically, from mother to a child, or horizontally via sexual contact, blood transfusion, etc. Observations show that children begin to increase HIV by sexual contact in Turkey. In this study, the aim is examining the transmission HIV dynamics in children in Turkey using a mathematical model.

SIR (Susceptible, Infected, Removed) mathematical modelling is used to determine the characteristics of infectious diseases. In this study, an SIR model was used with 3 equilibrium points. These are defined as no transmission, vertical and horizontal transmission routes denoted as the basic replication numbers R_1 and R_2, respectively. Equilibrium points' stability analysis was provided by using the Lyapunov function. If $R_1 < 1$ and $R_2 < 1$, vertical and horizontal transmission are not significant, if $R_1 < 1$ and $R_2 > 1$; the epidemic occurs via horizontal transmission, and for the condition $R_1 > 1$ and $R_2 < 1$, the epidemic occurs through the vertical transmission route.

Keywords Vertical transmission routes of pediatric HIV · Horizontal transmission routes of pediatric HIV · Mathematical model · Turkey

B. Kaymakamzade (✉) · F. T. Sa'ad · E. Hincal
Department of Mathematics, Near East University, Nicosia, Cyprus
e-mail: bilgen.kaymakamzade@neu.edu.tr

B. Kaymakamzade · T. Sanlidag · N. Sultanoglu · F. T. Sa'ad · M. Sayan · E. Hincal
DESAM Institute, Near East University, Nicosia, Cyprus

M. Sayan
Faculty of Medicine, Clinical Laboratory, PCR Unit, Kocaeli University, Kocaeli, Turkey

N. Sultanoglu
Faculty of Medicine, Department of Medical Microbiology and Clinical Microbiology, Near East University, Nicosia, Cyprus

D. Uzun Ozsahin et al. (eds.), *Decision Analysis Applied to the Field of Environmental Health*, Professional Practice in Earth Sciences,
https://doi.org/10.1007/978-3-030-96682-9_11

1 Introduction

The human immunodeficiency virus (HIV) can be transmitted via body fluids (e.g. blood, semen, pre-seminal fluid, vaginal fluids, rectal fluids and breast milk). Once the transmission is established, HIV attacks and destroys the immune system, especially the $CD4^+$ T-cells. Although HIV is a chronic disease that is manageable in the case of using antiretroviral therapy (ART), if not treated properly, it can lead to acquired immunodeficiency syndrome (AIDS), which is the advanced stage of the infection (Hınçal et al. 2018).

HIV is one of the leading infectious diseases which can be spread directly from human to human, but there are disparities in its prevalence. For understanding of disparities, it is important to understand the components and dynamics of the system. Mathematical modelling is one of the methods that plays an essential role in the transmission dynamics of HIV (Naresh and Sharma 2011). Some studies have focused on HIV transmissions, which have separated the transmission into vertical (from mother to child) (Kadi et al. 2014; Wang et al. 2010; Mugisha and Luboobi 2003; Gumel 2003; Waziri et al. 2012) and horizontal transmission (Hınçal et al. 2018; Children and HIV Fact Sheet 2016).

The transmission of HIV can be divided into two subgroups, which are called vertical and horizontal. Transmission through medical surgeries, dental processes, blood transmission, breastfeeding, HTC, bisexual intercourse, and men having sex with men (MSM) are considered as horizontal transmissions, whereas the transmission from mother to child during gestation and delivery is referred to as vertical transmission (Rubio and Gahona 2015).

According to the World Health Organization (WHO) statistics, the number of children infected with HIV under 15 years old is 1500. Ninety percent who are infected with HIV are from Africa. Since 2005, approximately 1,8 million (1,5–2 million) children around the world have been living with HIV. Around 160,000 (110,000–260,000) children, are newly infected with HIV and around 400 children are infected every day. However, 110,000 children passed away due to AIDS and related illnesses (Children and HIV fact sheet 2016; Global and statistics 2019; WHO 2006).

In this paper, a new model is constructed involving systems of ODE. Furthermore, global stabilities of equilibrium points are studied with the Lyapunov function. In 2016, with the obtained data from Turkey, numerical simulations are given to support the analytical results.

The organization of the paper is as follows: In Sect. 2, the model is presented with the equilibrium points and basic reproduction number. Also, the stability analysis of the equilibrium points is investigated in this section. In Sect. 3, results are given based on numerical simulations of the obtained data. In Sect. 4, the research findings are discussed. Finally, Conclusion is given in Sect. 5.

n 2: E_1 is globally asymptotically stable if $R_2 < 1$.

onsider the Lyapunov function

$$V(S, I_V, I_H) = S_1 g\left(\frac{S}{S_1}\right) + I_{V,1} g\left(\frac{I_V}{I_{V,1}}\right) + I_H, \tag{3}$$

, a positive function. It is clear that $I_H > 0$, which leads $V \geq 0$. We need to at $\dot{V} < 0$ holds. Taking the derivative of (3), we get,

$$\frac{S_1}{S}\Big)\dot{S} + \left(1 - \frac{I_{V,1}}{I_V}\right)\dot{I}_V + \dot{I}_H$$

$$\frac{S_1}{S}\Big)(\pi - \beta_1 I_V S - \beta_2 I_H S - \mu S) + \left(1 - \frac{I_{V,1}}{I_1}\right)(\beta_1 I_V S + (1-\varepsilon)\theta I_V - (v+\mu) I_V) + \beta_2 I_H S - (v+\mu) I_H$$

$$- \frac{\hat{S}}{S} - \frac{S}{\hat{S}}\Big) + I_H(R_2 - 1)$$

efore $\dot{V} < 0$ if $R_2 < 1$. Since the equilibrium exists when $R_1 > 1$, therefore obally asymptotically stable for the conditions $R_2 < 1$ and $R_1 > 1$.

m 3: $E_2$is globally asymptotically stable if $R_2 < 1$.

Consider the Lyapunov function

$$V(S, V_1, V_2, I_1, I_2) = S_2 g\left(\frac{S}{S_2}\right) + I_V + I_{H,2} g\left(\frac{I_H}{I_{H,2}}\right), \tag{4}$$

$g(x) = x - 1 - lnx$, a positive function. Since $I_V > 0$, $V \geq 0$. We need to at $\dot{V}$ is negative definite. Taking the derivative of (4), we get

$$\dot{V} = \left(1 - \frac{S_2}{S}\right)\dot{S} + \dot{I}_V + \left(1 - \frac{I_{H,2}}{I_H}\right)\dot{I}_2.$$

eaningful if $R_2 > 1$ it means that $S_2 < S_0$, then we have,

$$\left(1 - \frac{S_2}{S}\right)\dot{S} + \dot{I}_V + \left(1 - \frac{I_{H,2}}{I_H}\right)\dot{I}_2 < \mu\hat{S}\left(2 - \frac{S_2}{S} - \frac{S}{S_2}\right) + I_V(R_1 - 1).$$

re $\dot{V} < 0$ if $R_1 < 1$. Since E_1 exists when $R_2 > 1$, hence E_1 is globally otically stable if $R_1 < 1$ and $R_2 > 1$.

2 Methods

2.1 *Mathematical Model*

The population $N(t)$ is divided into three compartments, S, I_V, and I_H, which denotes the number of children in susceptible, HIV positive children with vertical transmission, HIV positive children with horizontal transmission, respectively.

2.1.1 Susceptible, S

Consider the recruitment rate π to the susceptible population per time t. As a result of infection or natural death (μ), susceptible individuals are removed from the compartment. Any susceptible child that contacted the infected individual can remove in susceptible class to infected class with horizontally with rate β_2 and β_3 and with rate β_1 susceptible children can be removed vertically.

2.1.2 HIV Positives with Vertically, I_V

Let ε be new-born HIV infected who die right after the birth and θ be new-born infected with HIV. Therefore, one of the input rate is $(1-\varepsilon)\theta$ that represents new-born children who are living and infected with HIV. Any Health children can be infected during breastfeeding with the rate of β_1. Any infected children according to vertical transmission can be removed in that class via AIDS (v) or natural death (μ).

2.1.3 HIV Positives with Horizontally, I_H

Whenever any susceptible children contacted with infected children can be infected with the rate $\beta_2(\beta_3)$. Any infected HIV children removed in that class via AIDS or natural death like I_V class.

The model for transmission dynamics in children is given by the system (1) by means of ordinary differential equations.

$$\begin{aligned}
\frac{ds}{dt} &= \pi - (\beta_1 + \beta_2 I_H + \beta_3 I_V)S - \mu S \\
\frac{dI_V}{dt} &= \beta_1 S + (1-\varepsilon)\theta I_V - (v+\mu) I_V \\
\frac{dI_H}{dt} &= (\beta_2 I_H + \beta_3 I_V)S - (v+\mu) I_H
\end{aligned} \tag{1}$$

Since,

$$\frac{dN}{dt} = \frac{ds}{dt} + \frac{dI_V}{dt} + \frac{dI_H}{dt}$$

the positive region of (1) can be obtained as

$$\Omega = \left\{(S, I_V, I_H) : S, I_V, I_H \geq 0, S + I_V + I_H < \frac{\pi}{\mu}\right\}.$$

2.2 *Equilibria Points*

For the evaluation of equilibrium points, right side of the system (1) is equated to 0. Solving the obtained system simultaneously, three equilibrium points are found as follows.

1. Disease Free Equilibrium Point;
   For the disease free equilibrium point, occurrence of no new infection is assumed, that is, β_1 is zero. So, the disease free equilibrium is,

   $$E_0 = (S_0, I_{V,0}, I_{H,0}),$$

   where, $S_0 = \frac{\pi}{\mu}$, $I_{V,0} = 0$, and $I_{H,0} = 0$. Since $\frac{\pi}{\mu} > 0$, E_0 always exists.
2. Vertical Transmission Endemic Equilibrium;
   If there is no horizontal transmission then $I_H = 0$, therefore the horizontal transmission endemic equilibrium is obtained as

   $$E_1 = (S_1, I_{V,1}, I_{H,1}),$$

   where, $S_1 = \frac{\pi}{\beta_1+\mu}$, $I_{V,1} = \frac{\pi\beta_1}{(\beta_1+\mu)[(1-\varepsilon)\theta-(\upsilon+\mu)]}$, and $I_{H,1} = 0$. For the existence of this equilibrium S_1, $I_{V,1}$, and $I_{H,1}$ must be positive. Therefore, it exists when $\frac{(1-\varepsilon)\theta}{\upsilon+\mu} > 1$.
3. Horizontal Transmission Endemic Equilibrium;
   If there is no horizontal transmission then $I_V = 0$ so $\beta_1 = 0$, therefore the horizontal transmission endemic equilibrium is obtained as

   $$E_2 = (S_2, I_{V,2}, I_{H,2})$$

   where, $S_2 = \frac{\upsilon+\mu}{\beta_2}$, $I_{V,2} = 0$, and $I_{H,2} = \frac{\pi}{\upsilon+\mu} - \frac{\mu}{\beta_2}$. For the existence of this equilibrium S_2, $I_{V,2}$, and $I_{H,2}$ must be positive. Therefore, it exists when $\frac{\beta_2\pi}{\mu(\upsilon+\mu)} > 1$.

2.3 *Basic Reproduction Number (BRN or*

Basic reproduction number is the number of secondary
infected person, in a fully susceptible population, whi
generation matrix method is used to find the basic rep
this method, the next generation matrix at E_0 is obtain

$$FV^{-1}(E_0) = \left[\frac{(1-\varepsilon)\theta}{\upsilon+\mu} 0 0 \frac{\beta_2}{\mu(\upsilon -}\right.$$

The spectral radius, which are the dominant eigenvalue
$\frac{\beta_2\pi}{\mu(\upsilon+\mu)}$. Therefore, the basic reproduction ratio is obtai

$$R_0 = \{R_1, R_2\},$$

where, $R_1 = \frac{(1-\varepsilon)\theta}{\upsilon+\mu}$, and $R_2 = \frac{\beta_2\pi}{\mu(\upsilon+\mu)}$.

2.4 *Global Stability of Equilibria*

In this subsection, the global properties of equilibrium
constructed a Lyapunov function to the show global s
point.

Theorem 1: The DFE E_0 is globally asymptotically sta

Proof: Consider the Lyapunov function,

$$V(S, I_V, I_H) = S_0 g\left(\frac{S}{S_0}\right) + I_v -$$

where, $g(x) = x - 1 - lnx$, which is a positive fu
and$I_2 > 0$, $V > 0$ and$V(S_0, I_{V,0}, I_{H,0}) = 0$.

For the global asymptotical stability of E_0, it is nece
Taking the derivative of (2), we obtain,

$$\dot{V} = \left(1 - \frac{S_0}{S}\right)\dot{S} + \dot{I}_V + \dot{I}_H$$

$$= (\pi + \mu S)\left(1 - \frac{S^0}{S}\right) + (\upsilon + \mu)I_V(R_1 - 1) +$$

From the boundedness of S, $S < S_0$, and hence $\dot{V} <$
Thus, E_0 is globally asymptotically stable for the conditi

Theore

Proof:

where,
show t

Theor

Proof:

where
show t

Therе
asym

3 Results

Kocaeli University, PCR Unit data of 69 vertical, and 152 horizontal, between 2010 and June 2019 [MSM; gay-men-children (n = 96), heterosexual-child (n = 39), bisexual-child (n = 8), blood transfusion (n = 6), medical/surgical procedure (n = 1) and tooth treatment (n = 1)] with a total of 221 HIV positive children were used as a data.

In this section, results are obtained by using the formulated data obtained from Kocaeli University, Clinical Laboratory polymerase Chain Reaction Unit in Turkey. Table 1 presents the parameter values which are calculated based on the data, obtained. Table 2 shows the values of basic reproduction number, based on the data obtained.

3.1 Equilibrium Points

With using the parameters in Table 2, we found the values of the equilibria as

1. Disease free Equilibrium Point,

$$E_0 = \left(S_0, I_{V,0}, I_{H,0}\right) = (9.539789, 0, 0).$$

2. Vertical transmission endemic equilibrium,

Table 1 Model parameters as calculated from the data

Parameters	Values
S	22, 883, 067
I_V	70
I_H	151
β_1	0.03
β_2	0.07
ε	0.11
θ	0.008
μ	0.2086
v	0.014
π	1.99

Table 2 Basic reproduction numbers

Vertical transmission	Horizontal transmission
$R_1 = \frac{(1-\varepsilon)\theta}{v+\mu}$	$R_2 = \frac{\beta_2\pi}{\mu(v+\mu)}$
1.614201578	3.257070842

$$E_1 = (S_1, I_{V,1}, I_{H,1}) = (5.2538558, 5.3344651, 0).$$

3. Horizontal transmission endemic equilibrium,

$$E_2 = (S_2, I_{V,2}, I_{H,2}) = (2.9289474, 0, 6.1950655).$$

3.2 Basic Reproduction Ratios (BRR or BRN)

3.3 Numerical Simulation

Simulating the above result, Fig. 1 shows the transmission dynamics of HIV/AIDS for children in Turkey. According to Fig. 1, horizontal transmission more dangerous than vertical transmission. Therefore, in Fig. 2 the dynamics of horizontal transmission is given.

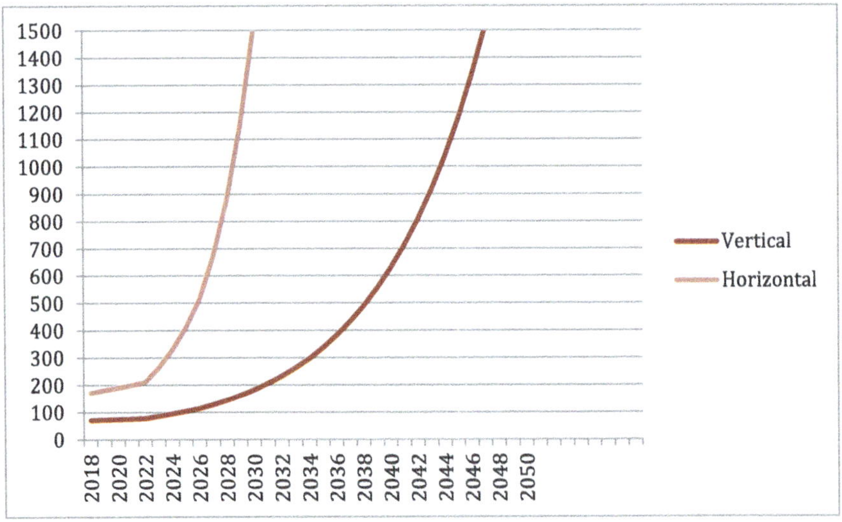

Fig. 1. The transmission dynamics of HIV/AIDS for children: Vertical line shows years, and horizontal line shows the number of people infected

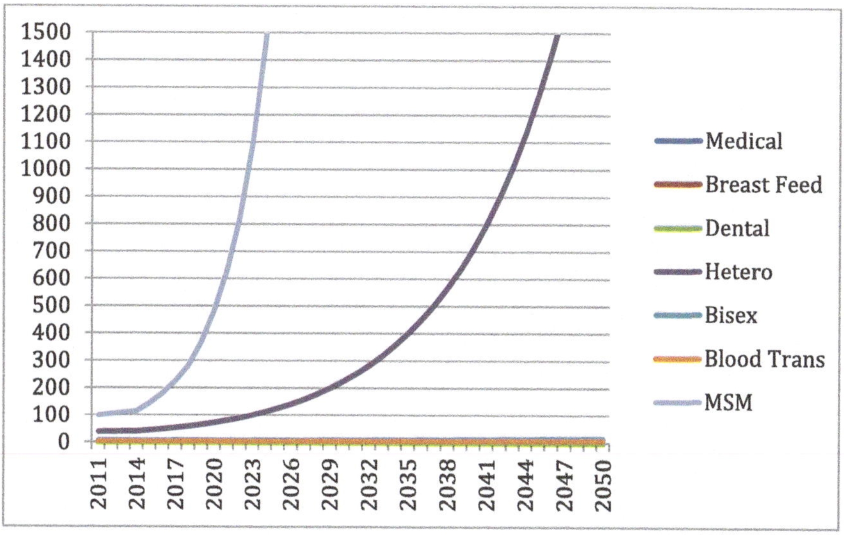

Fig. 2. Dynamics of Horizontal transmission: Vertical line shows years, and horizontal line shows the number of people infected

4 Discussion

A mathematical model was constructed to analyze the HIV transmission in children in Turkey. Three equilibrium points were found and with the use of Lyapunov function, stability of each equilibrium point is shown depending on the magnitude of the BRN. If the vertical and horizontal transmission routes of HIV infection, denoted by R_1 and R_2 respectively, are less than one, the disease-free equilibrium will be globally asymptotically stable. In the case of R_1 greater than one and R_2 less than one, the vertical transmission endemic equilibrium will be asymptotically stable. Also, if R_1 is less than one and R_2 is greater than one; horizontal transmission equilibrium will be globally asymptotically stable.

The population of Turkey in 2019 was 22, 883, 067 and the number of positive HIV children cases was 129. According to the Table 2, the BRR of vertical and horizontal transmission are 1.61 and 3.26, respectively. This implies that one positive HIV child in Turkey has the ability to transfer the disease to almost 2 children vertically and almost 3 children horizontally.

Numerical simulations are carried out; the results support the analytical findings. The simulation results predicted that within approximately 10 years, a total rise in HIV positive individuals with horizontal transmission of 1500 is expected, whereas the number of positive HIV individuals via vertical transmission is expected to rise by approximately 1500 cases in almost 50 years. This shows that horizontal transmission occurs more vigorously than vertical transmission. In Fig. 2, the simulation of horizontal transmission is shown. It can be observed that the HIV transmission in

MSM and heterosexual groups is increasing while the transmission via other routes is decreasing.

5 Conclusion

According to our mathematical model, it is concluded that both vertical and horizontal transmission routes of HIV infection of children in Turkey are at significant levels. However, if the increase in the MSM sexual contact route compared to the horizontal transmission is not addressed by the necessary authorities, in the next 30 years, HIV infections through MSM sexual contact will accelerate compared to other transmission routes.

References

Hincal E, Sayan M, Baba IA, Sanlidag T, Saad FT, Kaymakamzade B (2018) Dynamics of HIV-1 infected population acquired via different sexual contacts route: a case study of Turkey. Bull Karaganda Univ Math 91(3):83–88. https://doi.org/10.31489/2018M3/83-88

Naresh R, Sharma D (2011) An HIV/AIDS model with vertical transmission and time delay ISSN 1 746–7233. World J Model Simul 7:230–240

Kadi S, Itagimath SR, Gani SR (2014) Dynamic characteristic analysis of mother to child transmission of HIV in India. Global J Med Publ Health 3(6):1–7

Wang J, Reilly KH, Han H, Peng ZH, Wang N (2010) Dynamic characteristic analysis of HIV mother to child transmission in China. Biomed Environ Sci 23:402–408

Mugisha JYT, Luboobi LS (2003) Modeling the effect of vertical transmission in dynamics of HIV/AIDS in an age- structured population. S Pac J Nat Sci 21:82–90

Gumel AB (2003) Using mathematics to understand HIV pathogenesis and epidemiology. Book of Abstracts. In: African mathematics conference

Waziri AS, Massawe ES, Makinde OD (2012) Mathematical modelling of HIV/AIDS dynamics with treatment and vertical transmission. J Appl Math 2(3):77–89. https://doi.org/10.5923/j.am.20120203.06

Bashiru KA, Ojurongbe TA (2015) Stochastic analysis of heterosexual transmission of HIV/AIDS epidemic in the presence of treatment. J Nigeria Assoc Math Phys 31(1):27–34

Rubio EV, Gahona RG (2015) Vertical transmission of HIV—medical diagnosis. Therap Options Prev Strategy. https://doi.org/10.5772/61202

Children and HIV fact sheet (2016). https://www.unaids.org/sites/default/files/media_asset/FactSheet_Children_en.pdf. Accessed July 2016

Global HIV & AIDS statistics - 2019 fact sheet (2019). https://www.unaids.org/en/resources/fact-sheet

WHO: HIV in Children (2006). https://www.who.int/hiv/toronto2006/Children2_eng.pdf

Reliability of Covid-19 PCR Test Results with Statistical Distributions

Nezihal Gokbulut, Nazife Sultanoglu, Tamer Sanlidag, Murat Sayan, and Evren Hincal

Abstract COVID-19, the causative agent being severe acute respiratory syndrome coronavirus 2 (SARS-CoV-2) has been one of the most contagious diseases in the world leading to a pandamic. Many strategies have been improved and continue to be improving for the control the spread of the disease. Applying real-time reverse transcriptase polymerase chain reaction (RT-PCR) tests at certain intervals to individuals can be a control mechanism in the prevention of SARS-CoV-2 spread to others. In this study, a statistical model is applied to the positive SARS-CoV-2 cases obtained from NEU DESAM COVID-19 laboratory in the study period. The data is obtained from different type of kits which are Bio-Speedy® and DIAGNOVITAL® SARS-CoV-2 real time RT-PCR kits. The aim of this study is to investigate which kit is better for the control of the disease. For that purpose, R_t values of kits and the country are calculated separately via statistical model. As a result, it is concluded that with a little difference, DIAGNOVITAL®RT-PCR SARS-CoV-2 kit is better but both of the kits are suitable for the disease.

Keywords COVID-19 · Statistical modelling · Gamma distribution · PCR test kits

N. Gokbulut (✉) · E. Hincal
Department of Mathematics, Near East University, 99138 Nicosia, Cyprus
e-mail: nezihal.gokbulut@neu.edu.tr

N. Gokbulut · N. Sultanoglu · T. Sanlidag · M. Sayan · E. Hincal
DESAM Research Institute, Near East University, 99138 Nicosia, Cyprus

N. Sultanoglu
Faculty of Medicine, Department of Medical Microbiology and Clinical Microbiology, Near East University, 99138 Nicosia, Cyprus

M. Sayan
Faculty of Medicine, Clinical Laboratory, PCR Unit, Kocaeli University, Kocaeli 41001, Turkey

D. Uzun Ozsahin et al. (eds.), *Decision Analysis Applied to the Field of Environmental Health*, Professional Practice in Earth Sciences,
https://doi.org/10.1007/978-3-030-96682-9_12

1 Introduction

In late December 2019, unexplained acute respiratory disease outbreak emerged in Wuhan, China. Soon after, the causative agent was identified to be severe acute respiratory syndrome coronavirus 2 (SARS-CoV-2), which is a member of coronavirus family, leading to a disease referred to as COVID-19. It has been one of the most contagious diseases in a short time. SARS-CoV-2 has spread worldwide and evolved into a global pandemic resulting in millions of deaths (Wernike et al. 2020).

The common symptoms of the COVID-19 show similarities with the symptoms of Influenza virus (Molka et al. 2021). The transmission of the SARS-CoV-2 occurs through respiratory secretions of the infected person to others through indirect or direct transmission. Direct transmission involves inhaling the expelled respiratory droplets and/or aerosols of the infected person during coughing or sneezing whereas indirect transmission occurs through contaminated fomites via transporting the virus from fomites to mouth, nose or eyes (Transmission of SARS-CoV-2: implications for infection prevention precautions 2020).

The gold standard of the diagnosis of the SARS-CoV-2 from an infected person is by real-time reverse transcriptase polymerase chain reaction (RT-PCR) assay. To perform a real time RT-PCR samples are obtained from nasopharyngeal and oropharyngeal (throat) swabs. With the emerge of the COVID-19 pandemic many real time RT-PCR has been designed to detect SARS-CoV-2 from nasopharyngeal and oropharyngeal swab samples. Different SARS-CoV-2 real time RT-PCR kits detect different gene regions of the SARS-CoV-2. Screening the population for SARS-CoV-2 is key to identify and isolate the infected individuals to prevent further spread of the disease (Wernike et al. 2020).

With the rise of the COVID-19 pandemic many attempts have been made to study and analyze the disease characteristics. Epidemiology is a field of study that analyzes the structure of infectious diseases in specified populations. The role of mathematical and statistical modelling in epidemiology is crucial by means of control of the spread of diseases. Mathematical modelling divides the whole population into reasonable compartments and studies their relationship while statistical modelling applies suitable statistical distributions (e.g., gamma distribution, binomial distribution, etc.) to obtained data with the purpose of analyzing the current situation of the disease. The advantage of statistical modelling is that with right determined distributions, only a group of accurate data will be enough for the analysis to reveal good results (Huppert and Katriel 2013; Brauer 2017). Statistical modelling has been a method of control applied to epidemics and pandemics since the 1800s. Due to the use of a group of data in model construction, the validity of inferences depends on the quality of data in statistical modelling. Hence, the accuracy of data plays an important role, especially in health sciences (Siettos and Russo 2013; Why is Statistics Important in Decision-Making? 2021).

In mathematical modelling, basically, the whole population is divided into sensible compartments and the relationship between the compartments are expressed with ordinary differential equations. The basic reproduction number, R_0, is calculated, subsequently, to decide the infectiousness of the disease (Roberts and Heesterbeek 2012). In statistical modelling, the effective reproduction number, R_t, is calculated rather than the basic reproduction number. Different than R_0, in infectious diseases, R_t counts immune individuals as well. Also, R_t tracks the evolution of transmission which means that it can provide more information than R_0 (Kwok et al. 2020; Nishiura and Chowell 2009).

Mathematical and statistical modelling have been widely used and continues to be used for the control and analysis of the COVID-19 and other infectious diseases. (White and Enright 2010; Kar and Jana 2013; Wang et al. 2018) and (Hincal, Kaymakamzade and Gokbulut, Basic reproduction number and effective reproduction 2020) are some of the papers that examines the diseases via mathematical modelling with vaccine and control strategies while the papers (Siettos and Russo 2013) and (Hincal et al. 2021) are analyzing infectious diseases with the help of statistical modelling.

In this study, collected data in NEU DESAM COVID-19 laboratory is analyzed to see the effect and importance of different kinds of kits, and to decide their reliability. For this purpose, the improved statistical model in (Hincal et al. 2021) is modified according to the obtained data.

2 Methods

2.1 Laboratory Data Collection

The positive SARS-CoV-2 samples obtained from 1st of February to 16th of June 2021 in the NEU DESAM COVID-19 laboratory were used in this study. The positive diagnosed SARS-CoV-2 from the combination of oropharyngeal and nasopharyngeal swab samples via using the Bio-Speedy® and DIAGNOVITAL® for the same time period was compared using the methods in statistical modelling. For the comparison of kits with the data of country, daily cases are used that are taken from the Ministry of Health. In this study, daily cases of the country capture the dates 1st of February till the 16th of June, 2021.

The Bio-Speedy® SARS-CoV-2 Double Gene real time RT-PCR kit (Turkey, Version 19.04. 2021) targets the SARS-CoV-2 specific nucleocapsid (N) and ORF1ab genes. Also, as a control of the sampling, the kit detects human RNase P gene. For the February 2021 samples previous version of the kit was used. With the arise of the new version, the latest version was adapted in the NEU DESAM COVID-19 laboratory (Bioeksen 2021).

On the other hand, DIAGNOVITAL® SARS-CoV-2 real time RT-PCR kit v2.0 targets N1 and N2 regions of the Nucleocapsid genes of the SARS-CoV-2. In addition, the kit detects human RNase P gene as an internal control for ensuring if the sampling has been done successfully. This feature present in both of the kits used. Prior to release of v.2.0 of the kit, older version was used in the laboratory. With the release of the new version - v2.0 was adapted in the use of diagnosis of SARS-CoV-2 from the samples received from NEU Hospital.

Both of the mentioned kits and samples were used in the diagnosis of SARS-CoV-2 in NEU DESAM COVID-19 laboratory with the Insta Q96 plus Mumbai, India rRT-PCR device (DIAGNOVITAL SARS-CoV-2 Real-Time PCR Kit 2020). The only difference was the different kits being applied to the samples in the diagnosis of SARS-CoV-2. Thus, the present study aimed to compare the reliability of the two kits using statistical distributions.

2.2 Statistical Model

In this model different types of statistical distributions are used. Gamma distribution, Binomial distribution, and Posterior distribution are the selected distributions for the construction of the model.

First of all, for the prediction of incidence rate of cases, gamma distribution is applied. This distribution is chosen due to the delay of cases. So, the delay in reporting cases, δ, is

$$\delta \sim \Gamma(\mu, \sigma),$$

where μ is the mean of cases and σ is the standard deviation of cases. Afterwards, for the assumption of cases, denoted by C_t, Binomial distribution is used, which is known as n independent Bernoulli trials (Daniel 2005). In the study, PCR tests represented trials and the probability of having a positive result from the tests is assumed to be the success of Binomial distribution. Hence,

$$C_t \sim Binom\left(\int_0^{\infty} \Gamma(\mu, \sigma) I^r_{t-x} dx, r_\mu\right), \tag{1}$$

where r_t denotes the success of trials and I^r_{t-x} denotes the reported cases at time t.

The binomial formula in (1) is taken as a prior distribution and hence posterior distribution is applied for r_t. Here r_t that is mentioned above is the ratio of cases to reported cases at time t. Posterior distribution plays an important role for finding R_t values. In order to say that any infectious disease is under control with the help of R_t, the condition $R_t < 1$ should holds. If $R_t > 1$, then it is concluded as the disease continues to spread and hence, the conditions and restrictions are not enough yet.

The R_t values for the kits of NEU and Government are calculated, separately with the formula below:

$$R_t = \frac{FR}{r_t},$$

where FR denotes the future records of cases that are obtained above.

Table 1

From 1st February to 16th of June	The Country	DIAGNOVITAL®RT-PCR SARS-CoV-2 kit	The Bio-Speedy® SARS-CoV-2 real time RT-PCR kit
Median of R_t value with 50% Confidence Interval	1.02 (0.94–1.26)	1.00 (0.62–1.58)	1.10 (0.74–1.62)

At this step, calculations are made by using the previous week's data. While constructing the model, incubation period of the COVID-19 is also taken into account with gamma distribution, i.e., $\Gamma(x|\mu, \sigma) > 0$ for $x \in \{1, 7\}$. With this assumption as a null hypothesis scenario, contribution to rate of cases will be very small and the results of the study will not be affected.

With the help of this model, daily median of R_t values are calculated and for the accurate results and comparison, 50% of confidence interval are taken. The results are shown in Table 1.

3 Conclusion

Positive SARS-CoV-2 cases results obtained in NEU DESAM COVID-19 laboratory obtained via using two separate real time SARS-CoV-2 kits which were Bio-Speedy® and DIAGNOVITAL® were compared via the statistical model. The only difference between the comparison were the kits and nothing else as same infectious disease experts and vehicle was used to perform SARS-CoV-2 real time RT-PCR assays.

The comparison revealed that both of the kits have indicated same pattern as shown in Table 1. This suggests that since both of the kits are indicating the same pattern the results are consisted. Thus, we suggest that both of the kits used are reliable in the diagnosis of SARS-CoV-2. However, the median of R_t value of the DIAGNOVITAL®RT-PCR SARS-CoV-2 kit is closer to the R_t value of the country when it is compared with the results o Bio-Speedy® real time RT-PCR kits.

If there was dramatic difference between the compared two kits then this was assumed to be a not reliable result. The samples were obtained from the people leaving in the same population hence consistent results were expected. Otherwise, a concern between the used kits could have arisen, questioning if one is better than the other in means of detecting the SARs-CoV-2 efficiently. However, this is not the case in this comparison since both of the kits indicated very similar patters suggesting that both is efficient and reliable in terms of detecting the SARS-CoV-2 from obtained samples. It is worth to mentioned that not only kits but also laboratory staff running the real time RT-PCR assays is important. An unexperienced laboratory staff may lead to unreliable results. As a conclusion of this study we declare that both of the kits used DIAGNOVITAL® and Bio-Speedy® is reliable in the diagnosis of SARS-CoV-2.

References

Bio-Speedy® Direct RT-qPCR SARS-CoV-2 (24 May 2021) Bioeksen R&D Technologies Inc. https://www.fda.gov/media/141823/download

Brauer F (2017) Mathematical epidemiology: past, present, and future. Infect Dis Model 113–127

Daniel WW (2005) Biostatistics. Wiley, USA

DIAGNOVITAL SARS-CoV-2 Real-Time PCR Kit (5 June 2020) https://www.fda.gov/media/138928/download

Hincal E, Kaymakamzade B, Gokbulut N (2020) Basic reproduction number and effective reproduction. Bull Karaganda Univ 86–95. https://doi.org/10.31489/2020M3/86-95

Hincal E, Kaymakamzade B, Suren N, Gokbulut N (2021) Estimating Covid-19 deaths by using binomial model. In: AIP conference proceedings, vol 2325, no. 1, p 020050

Huppert A, Katriel G (2013). Mathematical modelling and prediction in infectious disease epidemiology. Clin Microbiol Infect 999–1005

Kar TK, Jana S (2013) A theoretical study on mathematical modelling of an infectious disease with application of optimal control. Biosystems 111(1):37–50

Kwok KO, Lai F, Wei WI, Wong SY, Tang JW (2020) Herd immunity – estimating the level required to halt the COVID-19 epidemics in affected countries. J Infect 80(6):e32–e33

Molka O, Klopfenstein T, Belfeki N, Gendrin V, Zayet S (2021) A comparative systematic review of COVID-19 and influenza. Viruses 13(3):452

Nishiura H, Chowell G (2009) The effective reproduction number as a prelude to statistical estimation of time-dependent epidemic trends. Math Stat Estim Approaches Epidemiol 103–121

Roberts MG, Heesterbeek JA (2012) Characterizing the next-generation matrix and basic reproduction number in ecological epidemiology. J Math Biol 66(4–5):1045–1064. https://doi.org/10.1007/s00285-012-0602-1

Siettos CI, Russo L (2013) Mathematical modeling of infectious disease dynamics. Virulence 4(4):295–306

Transmission of SARS-CoV-2: implications for infection prevention precautions (9 July 2020) World Health Organization. https://www.who.int/news-room/commentaries/detail/transmission-of-sars-cov-2-implications-for-infection-prevention-precautions

Wang Z, Zhao D, Wang L, Sun G, Jin Z (2018) Two-strain epidemic model with two vaccinations. Chaos Solitons Fractals 106:342–348

Wernike K, Keller M, Conraths FJ, Mettenleiter TC, Groschup MH, Beer M (2020) Pitfalls in SARS-CoV-2 PCR diagnostics. Transbound Emerg Dis 68(2):253–257

White PJ, Enright MC (2010) Mathematical models in infectious disease epidemiology. Infect Dis 70–75. https://doi.org/10.1016/B978-0-323-04579-7.00005-8

Why is Statistics Important in Decision-Making? (5 April 2021) Michigan Tech https://onlinedegrees.mtu.edu/news/why-statistics-important-decision-making

Performance Evaluation of the Petrol Production Methods in Bakken Reservoirs

Fondjo Fondjo Yann Muriel, Dilber Uzun Ozsahin, and Berna Uzun

Abstract Bakken reserves are non-conventional reserves which englobes a huge source of raw material in the world as a whole due to the constant extraction of conventional oil reserves with time leading our attention on these particular ones. This study aims at making a synthetic analysis of the various petroleum production methods through documentation and a mathematical simulation using Multi Criteria Decision Making (MCDM) Fuzzy-Preference ranking organization method for enrichment evaluation (F-PROMETHEE) method to determine the appropriate one in bakken reservoir. In this optic we had to bring out the different variables table into consideration for the determination of the functionality and efficiency of a particular method during production in bakken reservoirs.

Through this research we found that during production in our reservoir the drive mechanisms couldn't provide effective productivity with time and determining which

F. F. Y. Muriel
Faculty of Engineering, Department of Petroleum Engineering, Near East University, Nicosia, TRNC, Mersin 10, Turkey

Faculty of Civil and Environmental Engineering, Department of Civil Engineering, Near East University, Nicosia, TRNC, Mersin 10, Turkey

D. U. Ozsahin
Medical Diagnostic Imaging Department, University of Sharjah, College of Health Science, Sharjah, United Arab Emirates
e-mail: dozsahin@sharjah.ac.ae

D. U. Ozsahin · B. Uzun (✉)
Center of Operational Research in Healthcare, Near East University, Nicosia, TRNC, Mersin 10, Turkey
e-mail: berna.uzun@neu.edu.tr

D. U. Ozsahin
Faculty of Engineering, Department of Biomedical Engineering, Near East University, Nicosia, TRNC, Mersin 10, Turkey

B. Uzun
Faculty of Arts and Sciences, Department of Mathematics, Near East University, Nicosia, TRNC, Mersin 10, Turkey

Department of Statistics, Carlos III University of Madrid, 28903 Getafe, Spain

D. Uzun Ozsahin et al. (eds.), *Decision Analysis Applied to the Field of Environmental Health*, Professional Practice in Earth Sciences,
https://doi.org/10.1007/978-3-030-96682-9_13

enhanced oil recovery method to use was sometimes complex due to the multitude of criteria's taken into considerations which draws our attention on the F-PROMETHE which is used for MCDM problems in many other fields. Using these method, the variables have been identified and graded in a certain range to ease input and understanding. They have been processed depending on the impact of each criterion and the results using this method can be modified if the variables are been changed. The results showed that the bakken reservoir technique is the most effective technique based on the parameters used (such as; lithology, fracture, depth, porosity, permeability, etc.) followed by HC gas injection and water injection. The F-PROMETHE method is efficient for the petroleum sector as in other fields where it has been used for MCDM problems.

Keywords Oil · Petrol production methods · Bakken reserves · Decision analysis · Fuzzy logic

1 Introduction

Oil, mainly known as "black gold", is the major component of liquid mining and is a major factor in a country's economy. The petroleum sector is subdivided into three main segments: we have the up stream for exploration and production, the middle stream for the processing and transport of crude oil, and then the down stream for the distribution and marketing of hydrocarbon products. My research had for main objective, making a synthetic analysis through detailed documentation of recovery methods and their effectiveness in a bakken reservoir, their recovery factors and importance in the productivity of a reservoir. In this, we had to examine the nature and parameters of the reservoir, identify the drive mechanisms, which are found in there. This was to have a brief overview on the types of fluid and its productivity.

The oil field productivity depends on the reservoir size, complexity, primary mechanisms, type and quantity of fluids it contains. Hence, it is essential to define the formation characteristics and prediction of the production with the view to establish the feasibility in the oil exploration (John 2003). For this reason, reservoir simulation is widely used to analyze the reservoir behavior, set and optimize a development plan of increasing the oil recovery efficiency. The production of hydrocarbons in natural depletion reservoir is accompanied by different mechanisms, which are; compressibility of the rock and fluids, expansion of the gas cap, dissolved gas and drainage by aquifer (Elhadi 2016). Considering the increasing reduction of oil resources in the world, it is necessary to use recovery methods and enhanced oil recovery from reservoirs because during production, when the gas cap expands and the dissolved gas is evolved from solution there is a decrease in the reservoir's energy (Enginius 2013).

Bakken reservoirs are mainly shaly geologic formations which contains mainly non-conventional gas [oil and gas mainly from biogenic origins]. Oil and gas were primarily found in this formation in 1951 which are mainly produced by hydraulic fracturing to increase the pore spaces in the rock so as the oil and gas should be released and can be produced easily (CBC 2008). The extraction of petroleum is the process by which usable hydrocarbons are drown out from beneath the earth surface through a series of recovery methods which can be listed as follows: primary, secondary, tertiary, and infill recovery (Glover 2013). This stage can also be called the Kareem stage. This is the recovery of hydrocarbons from the reservoir using the natural energy of the reservoir as a drive. The reservoir drive comes from a number of natural mechanisms with different energy sources (McCain 1990). Early in the history of a reservoir the drive mechanism will not be known, it is determined by analysis of production data [reservoir pressure and fluids production ratio]. The earliest possible determination of the drive mechanism is a primary goal in the early life of the reservoir as its knowledge can greatly improve the management and recovery of reserves from the reservoir in its middle and later life (cbcTotal 2007). The types of drive here are five; solution gas cap, gas cap, water drive, gravity drainage, combination and mixed drive (Saifaddeen Sallam 2014).

During the production when the well is no more capable to produce with its natural energy, we move the secondary methods to stimulate and increase production rate. The main methods found under the secondary category are water injection and air injection, HC gas injection. This recovery method is a result of human intervention in the reservoir to improve recovery when the natural drives have diminished to unreasonably low efficiencies because overtime, the lifetime if a well decreases with pressure falls (Elhadi 2016). At some points there is insufficient underground pressure to pull the oil to the surface. This recovery method relies in supplying the external energy to the reservoir by injecting fluid to increase reservoir pressure. This method increases the mobility of oil in order to increase extraction. Tertiary recovery begins when secondary oil recovery is not enough to continue adequate extraction, but only when the oil can still be extract profitably. This depends on the cost of the extraction method and the current price of crude oil. When price is high, previously unprofitable well are brought back in to use and when they are low, extraction is curtailed (Glover 2013). Primary and secondary methods usually recover only about 35% of the original oil in place. Many enhance oil recovery methods have been designed to do this and a few will be review here. They fall into three categories; thermal, chemical and miscible gas. These methods are extremely expensive for usage and implemented after expensive scale studies have isolated the reservoir rock characteristics that are causing oil to remain unproduced by conventional methods (Schlumberger 2011). The thermals are steam injection, fire flooding and steam injection. The chemical injections are polymer injection, microbial injection, foam injection and the last group we have miscible gas, which are mainly co2 injection, air injection, liquid carbon dioxide superfluid's and water alternating gas.

During the production phase in a well the production rate is been determined depending on the different types of drive mechanisms found in the reservoir. When the initial energy found in the well diminishes, an increase in the rate of production is been carried based on the compatibility with the lithology, the formation fluid and many other parameters but generally at this point the secondary recovery takes which is mainly HC injection and water injection or water flooding.

Water flooding is inexpensive and simple to use. Thus, it is dominant among fluid injection methods and is without question responsible for the current high oil producing rate within the America and Canada (Amir Amiri 2017). Water injection is a technique that uses injection wells and production wells to avail the energy with the water injected for displace oil to the production wells. In water flooding, water would displace oil from pores in a formation in a manner representing a leaky piston. In addition, the higher viscosity of oil in comparison to water will contribute to non—deal displacement behavior, several, concepts will be defined in order that an understanding of displacement efficiencies can be achieved (McCain 1990; Kaïchun 2017). As been mentioned above, water injection is the most important assistant recovery method for a reservoir due to its high recovery factor which is between 35 and 45% as mentioned earlier. The treatment of water before injection has for objective to remove suspended solids, dissolved minerals (Bill Bailey 2007).

Gas flooding method is similar to water flooding in principal and is used to maintain gas cap pressure even. In this method basically the gas produced in the early stage of the well during the primary production with the aid of drive mechanisms is been pumped in the same reservoir from another well to maintain the pressure of the gas cap. The incremental recovery ranges from 15 to 25% (Schlumberger 2011).

This method increases the mobility of oil in order to increase extraction. Tertiary recovery begins when secondary oil recovery is not enough to continue adequate extraction, but only when the oil can still be extract profitably. This depends on the cost of the extraction method and the current price of crude oil. When prices are high, previous unprofitable well are brought back in to use and when they are low, extraction is curtailed (Glover 2013). The tertiary methods are applied to all types of oil and gas reserves even though they are more effective on heavy oils such as polymer. Polymer flooding is an enhanced oil recovery method which functions as follows polymers which is a mixture of chemicals which can dissolve the hydrocarbons and increase its mobility by increasing the viscosity of the displacing water thereby raising the production rate.

Microbial enhanced oil recovery is a type of enhanced oil recovery which is based on biological technology by operating with the microorganisms found in the reserves (McCain 1990). Through this method an increase in mobility of the oil is been experienced because it helps sweep the oil trapped in porous medium with the aid of stimulants and indigenous microbes which increases economic benefit (CBC 2008). These particular methods compared to other enhanced recovery methods is less expensive to the company due to the fact that it deals with the natural environment and the microbial organisms found in there. This method generally sweeps up to 2/3 of the original oil in place found in the reservoir (Water injection operations and gas-injection sensitivities in the bakken formation volume 2017).

Chemical injection on its own is mainly all types of recovery method which increase the oil recovery by the injection of hydrocarbon-based chemicals which increase the mobility of the oil by dissolving it, reducing its corrosion and upgrade crude oil (BP statistical review of world energy 2015).

Foam injection is an enhanced recovery method which functions with the injection of foam in our reserve. This method operates on the basic of deliquificaton also referred to as "gas well dewatering", is the general term for technologies used to remove water or condensates build-up from producing gas wells (She et al. 2019).

This technic is more effective in gas well than oil well due to the high presence of gas molecules and the concentration of the gas cap.

Fire flooding and steam injection basically function the same way at a little difference due to the fact that their operating system is based on an increase in temperature in the reservoir to reduce the viscosity of the hydrocarbons and increase their mobility and quantity (Bill Bailey 2007). This by so doing causes more upward movement of the hydrocarbons in to the wellbore and renders the drive mechanisms more effective. In steam injection, the steam can be injected into the well directly from the producing well or from another and then left for a month for heating. These methods help in up to 50% of the production in heavy oil reserves (Schlumberger 2011).

Co2 injection this method can be identified as a miscible enhanced oil recovery method and another alternative gas can be used which is nitrogen. They are mainly used in reservoirs with very low permeability and they reduce deposition pressure in the reservoir and asphaltene deposit increases in the reserves (She et al. 2019). This method is highly costly and increases contamination in our reservoir.

Liquid carbon dioxide superfluid's in this particular case the carbon dioxide is been injected in the well and will dissolve in the oil causing the oil to swell and a reduction in the surface tension is been observed and part of the carbon dioxide is been produced in the oil along with the oil (Lea et al. 2003). This method is mainly effective in reservoirs deeper than 2000ft. the oil viscosity reduces and the carbon dioxide produced can be reinjected thus reducing the cos of production.

Water alternating gas injection simply is the injection of both CO2 and water at the same time with the gas increasing the volume of the oil by dissolving in it and the water moving in to the aquifer due to density difference between the two fluids.

In this study F-PROMETHEE method will be applied for the evaluation of the petrol production methods in bakken reservoirs.

2 Methodology

Multi-criterion analysis is a decision-making tool developed to solve complex multi-criteria problems that include several qualitative and quantitative aspects in the decision-making process (Uzun Ozsahin et al. 2020). The field of multi-criterion optimization is undergoing a significant evolution. This has resulted in the development of a large number of multi-criteria methods.

In real-life problems, information about systems is not always clear, often involving uncertain conditions (Ozsahin et al. 2020). In order to define such conditions in a simple form mathematically, Zadeh has been introduced the fuzzy sets, which is an expansion of the Boolean logic, in order to define the non-precise conditions mathematically (Ozsahin et al. 2018). This logic enables the decision makers to analyse the vague conditions analytically after the defuzzification process.

One of the effective MCDM method is PROMETHEE method which contains a pairwise comparison between the alternatives.

Table 1 Triangular linguistic fuzzy scale

Linguistic scale for evaluation	Triangular fuzzy scale
Very high [VH]	0.75, 1, 1
Important [H]	0.5, 0.75, 1
Aesthetics medium [M]	0.25, 0.50, 0.75
Low [L]	0, 0.25, 0.5
Very low [VL]	0, 0, 0.25

PROMETHEE I method gives a partial preorder of the options which contains the situations where the incomparability occurs in some cases. PROMETHEE II gives the net ranking of the options based on the net flows of the options (Ozsahin et al. 2018).

PROMETHEE methods are based on the following three steps (Sarigul Yildirim et al. 2021):

1- Enrichment of the preferred structure: We define a preference function. This allows for the magnitudes of the differences between evaluations on the different criteria to be taken into account.
2- Enrichment of the dominance relationship: An upgrade value relationship taking into account all the criteria are proposed and for each pair of shares, a degree of overall preference of one action over the other will be established.
3- Decision support: The upgrade relationship is used to inform the decision maker. PROMETHEE I will provide partial order of options, while PROMETHEE II provides net ranking.

In this study we defined the parameters of the petrol production methods in bakken reservoirs using the linguistic fuzzy scale as seen in Table 1 in order to construct the decision matrix.

Then we obtained the decision matrix of the petrol production methods in bakken reservoirs as shown in Table 2. Using the Yager index the fuzzy numbers assigned to linguistic conditions have been defuzzified. Then the Gaussian preference function have been selected as the preference function the each criterion for analysis of the PROMETHEE technique. The importance weights of the parameters have been defined by using fuzzy scale and shown in the decision matrix based on expert opinion.

3 Results

From the results obtained through our mathematical simulation using the F-PROMETHEE method the above data illustrates the different production methods based on a several criteria with sample a bakken reservoir. Through our analysis we found that bakken reservoirs are the most effective reservoirs with non-conventional hydrocarbons which means they are reserve found poor accessible zones may be due

Table 2 Parameters of the petrol production methods in bakken reservoirs

EOR method	Lithology	Fracture	Depth [ft]	porosity	Permeability [mD]	Reservoir thickness [ft]	Oil API gravity	Oil viscousity [cp]	Oil saturation	Reservour pressure [psia]	Resevoir temperature [F]
Importance weights	VH	M	VH	VH	VH	M	VH	VH	VH	VH	VH
Aim	Max	Min	Max	Max	Max	Min	Max	Min	Max	Max	Min
Air injection	Medium	No	Medium	Good	Bad	Low	Medium	Low	Good	Medium	Low
co2 injection	Good	Yes	Medium	Good	Bad	Low	Medium	Very low	Bad	Very low	Low
HC gas injection	Good	No	High	Good	Bad	Low	High	Low	Good	Good	Low
Chemical injection	Good	No	Very high	Medium	Bad	Medium	Very high	Medium	Very good	Medium	Medium
Thermal method [Steam injection]	Medium	No	Medium	Medium	Very good	High	Very low	Very high	Very good	Very high	Very high
Foam injection	Good	No	Medium	Very good	Good	Medium	High	High	Good	Medium	Low
Water injection	Very good	No	High	Very good	Very good	Very high	Medium	Medium	Good	Medium	Medium
Bakken reservoir	Very good	Yes	Very high	Very good	Good	Very high	Very high	Medium	Very good	Very high	Medium
Steam flooding	Good	No	High	Good	Very bad	Low	Very low	Very high	Very good	Very high	High
Microbial injection	Good	No	Medium	Bad	Good	Medium	High	Very high	Good	Low	High

(continued)

Table 2 (continued)

EOR method	Lithology	Fracture	Depth [ft]	porosity	Permeability [mD]	Reservoir thickness [ft]	Oil API gravity	Oil viscousity [cp]	Oil saturation	Reservour pressure [psia]	Resevoir temperature [F]
Polymer flooding	Good	No	Medium	Good	Good	Medium	Very high	Very high	Good	Medium	Medium
Liquid carbondioxide super fluids	Medium	No	High	Very good	Bad	Medium	Medium	Low	Bad	Very low	Low
Water alternating gas	Good	No	High	Medium	Good	High	High	Low	Bad	Very low	High
Fire flooding	Good	No	Medium	Bad	Very bad	Very low	High	Low	Very good	Very high	Very high

Table 3 The ranking results of the petrol production methods in bakken reservoirs with F-PROMETHEE

Rank	Alternatives	Net flow	Positive outranking flow	Negative outranking flow
1	Bakken reservoir	0.0883	0.1110	0.0227
2	HC gas injection	0.0450	0.0675	0.0225
3	Water injection	0.0384	0.0736	0.0352
4	Foam injection	0.0309	0.0610	0.0301
5	Chemical injection	0.0260	0.0625	0.0364
6	Air Injection	0.0882	0.0511	0.0429
7	Polymer flooding	0.0074	0.0526	0.0452
8	Fire flooding	−0.0036	0.0732	0.0768
9	CO2 injection	−0.0286	0.0552	0.0838
10	Liquid carbondioxide super fluids	−0.0341	0.0489	0.0830
11	Water alternating gas	−0.0380	0.0427	0.0807
12	Thermal method	−0.0437	0.0637	0.1074
13	Steam flooding	−0.0462	0.0517	0.0978
14	Microbial injection	−0.0502	0.0317	0.0819

to the presence of geological features such as faults and many other this particular reserves mainly contain heavy oil and gas (see in Table 3). Their poor accessibility only gives room for methods of penetration hydraulic fracturing of the rock, which increases the voids of the formation. The most effective petrol production methods in bakken reservoirs are obtained as the bakken reservoir, followed by HC gas injection and water injection and the least effective ones are obtained as the thermal method, steam flooding and microbial injection based on their net ranking values.

As shown in Fig. 1, the advantages of the petrol production methods in bakken reservoirs are obtained as the parameters above the 0-threshold, while the disadvantages of them are have obtained as the parameters below the 0-threshold.

4 Discussion

In this section of our work, we will focus on the results obtained and the method used to transform the data base in to that schematic result. Firstly, the F-PROMETHEE method is kind of a mathematical simulation method in which different parameters are been introduced for multicriteria determination for different methods to identify the best for a particular purpose. These criteria are been classified using simple and comprehensible inputs such as medium, good, bad, yes or no. these simplifies comprehension for the audience and the criteria are been classified based on their

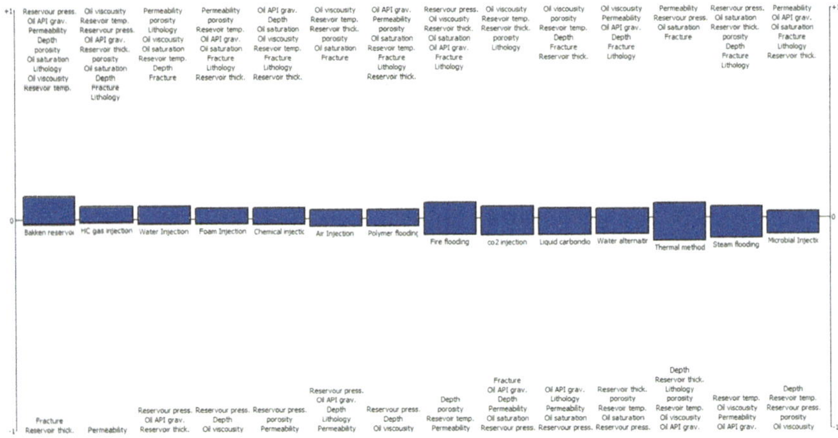

Fig. 1 F-PROMETHEE evaluations results of the petrol production methods in bakken reservoirs

importance and impact in the different methods. This method does not only function in a linear pattern but also takes in to consideration the changing of any variables which changes the output by so doing causing modification in the results.

The F-PROMETHEE method was used in this project to identify the adequate method for production in bakken reservoirs in the petroleum sector. Through this technique we could identify that thermal methods are the most appropriate in the production of hydrocarbon from bakken reserves which is extremely important due to the great reduction in oil reserves which gives room for the production of non-conventional reserves which represents a high percentage of the oil reserves due to the increase in demand and the diminishing quantity of conventional reserves. This brings us to making a detailed study through simulation of the appropriate method of production mainly in bakken reserves. Due to the fact that the F-PROMETHEE process classify the methods in terms of their importance or impact based on the effect they have on the different variables (criteria) we came to the conclusion that the most adequate method for the production of hydrocarbons from bakken reserves are thermal methods.

It should be understood that every method mentioned earlier in the results can be used in the extraction of hydrocarbon but based on the different criteria there are variations such as the types of hydrocarbon found, the API of the oil, the viscosity, permeability, porosity and many others. Through this method the variables give us access to any change in formation parameters based on the properties of every methods used. Thus, resulting in the best method used for a particular project or experiment.

5 Conclusion

We can conclude with assurance that the F-PROMETHEE methods gives us adequate and proper results with easy understandable inputs and variables resulting in an end product with a high operational probability. With the aid of this process, we could determine the best extraction methods of nonconventional hydrocarbons in bakken reservoir with a multitude of criteria, which can be easily adjusted in case there is a change in the primary variables.

References

John FC (2003) Hydrocarbon exploration and production developments in petrolium science, 7th edn. United Kingdom, p 25

Elhadi AM (2016) Analyse de performance d'injection d'eau dans la zone 4 du champ de Hassi Messaeud, p 40

Enginius (2013) Water injection course for petrolium engineering, p 60

CBC (June 2008) Bakken formations will it fuel Canada oil industry cbc

Glover DP (2013) Formation evaluation. Master course, p 44

McCain WD (1990) The properlies of petroleum fluids. Penn wellboods, p 11

cbcTotal (2007) Exploration et production dans le processus d'injection d'eau, 61

Saifaddeen Sallam MM (2014) The effect of water injection on oil well productivity, 34, San Antonio, Texas, USA

Schlumberger (2011) Schlumberger simulation launcher, 17

Amir Amiri AK (2017) Assessment the effect of water injection on improving oil recovery in X field. Published in Abadan, p 50

Kaïchun Yu KQ (2017) A method to calculate reasonables water injection rate for oil field, p 45

Bill Bailey MC (2007) Water control, p 10

Water injection operations and gas-injection sensitivities in the bakken formation volume 69 (01 October 2017)

BP statistical review of world energy 2015 British petroleum (June 2015)

She H et al (2019) Recent advance of microbial enhanced oil recovery in China

Lea JF, Nickens HV, Wells MR (2003) Gas well deliquification solutions to gas well liquid loading problems. Gulf professional publishing

Uzun Ozsahin D, Uzun B, Ozsahin I, Taiwo MM, Sani Musa M (2020) Fuzzy logic in medicine. In: Zgallai W (ed) Developments in biomedical engineering and bioelectronics, biomedical signal processing and artificial intelligence in healthcare. Academic Press, pp 153–182. Chapter 6

Ozsahin DU, Gichamo T, Gokcekus H, Gelete G, Uzun B (2020) Evaluation of different natural wastewater treatment alternatives by fuzzy PROMETHEE method. Desalin Water Treat 177(2020):400–407. https://doi.org/10.5004/dwt.2020.25049

Ozsahin DU, Isa NA, Uzun B, Ozsahin I (2018) Effective analysis of image reconstruction algorithms in nuclear medicine using fuzzy PROMETHEE. IEEE. https://doi.org/10.1109/ICASET.2018.8376892

Ozsahin I, Uzun B, Isa NA, Mok GSP, Ozsahin DU (2018) Comparative analysis of the common scintillation crystals used in nuclear medicine imaging devices. In: IEEE 60 nuclear science symposium and medical imaging conference proceedings [NSS/MIC]. https://doi.org/10.1109/NSSMIC.2018.8824485

Sarigul Yildirim F, Sayan M, Sanlidag T, Uzun B, Uzun Ozsahin D, Ozsahin I (2021) Comparative evaluation of the treatment of COVID-19 with multicriteria decision-making techniques. J Healthc Eng Article ID 8864522, 11 pages. https://doi.org/10.1155/2021/8864522

Socio-spatial Interactions Within Modern Workspace Interiors Post Covid-19

Shrouq Altamimi, Zeynep Üstün Onur, Dilber Uzun Ozsahin, and Berna Uzun

Abstract Though pandemics and contagious diseases have been an impending danger to humans since antiquity, the COVID-19 pandemic revealed a near-complete lack of readiness to combat such incidents in terms of built urban environment design. Workplaces, which are vibrant centers for daily human activity globally, proved to be active hubs for transmitting COVID-19 early on. This necessitates serious reconsideration of typical workspace design to sustainably allow workers back into the space without compromising public safety. Not only is this crucial during the ongoing COVID-19 pandemic, but solidification of design requirements that take this aspect into account is of paramount importance for effectively combating future pandemics. This paper investigates available research and advancements on design considerations and mitigation measures to be adopted when designing a pandemic-ready workspace.

S. Altamimi (✉)
Department of Interior Design, Near East University, Nicosia, Turkish Republic of Northern Cyprus, Turkey

Z. Ü. Onur
Department of Architecture, Near East University, P.O. Box: 99138, Nicosia, Mersin 10, TRNC, Turkey
e-mail: zeynep.onur@neu.edu.tr

D. U. Ozsahin
College of Health Science, Medical Diagnostic Imaging Department, University of Sharjah, Sharjah, United Arab Emirates
e-mail: dozsahin@sharjah.ac.ae

Department of Biomedical Engineering, Near East University, P.O. Box: 99138, Nicosia, Mersin 10, TRNC, Turkey

D. U. Ozsahin · B. Uzun
Center of Operational Research in Healthcare, Near East University, P.O. Box: 99138, Nicosia, Mersin 10, TRNC, Turkey
e-mail: berna.uzun@neu.edu.tr

B. Uzun
Department of Mathematics, Near East University, 99138, Nicosia, Mersin 10, TRNC, Turkey

Department of Statistics, Carlos III University of Madrid, 28903 Getafe, Spain

D. Uzun Ozsahin et al. (eds.), *Decision Analysis Applied to the Field of Environmental Health*, Professional Practice in Earth Sciences,
https://doi.org/10.1007/978-3-030-96682-9_14

Measures studies range from conscious socio-spatial zoning to provisions to ensure adequate hygiene and directly curb transmission.

Keywords Post Covid-19 pandemic office workspace design · Workplace interior design post covid-19

1 Introduction

Interior designers have a moral responsibility to society and public health since building and decorating spaces has a significant effect on overall health and well-being. When it comes to workplace design, this is quite essential. The amount of time spent in these settings as well as the difficulties individuals will face while there plays a major role in our lives. It is important that health and well-being will not be inhibited in these environments, but rather nourished and invigorated to allow individuals to function at their best. These environments must be designed with the specific requirements of those who will use them at the core of the design process, which must be supported by scientific research to obtain the best results. To put it another way, while planning office layouts, one must use an evidence-based and human-centered approach. Following the COVID-19 pandemic, the workplace design industry has faced a new difficulty. The user's safety must now be placed at the forefront of workplace design process. More essential than ever, designers need to adopt a human-centered approach and be fully aware of the implications of their choices for people who use the spaces they design. This pandemic has been one of the most transformative events since the beginning of the twentieth century. A lot of people believed that this nightmare would have been over by now. The World Health Organization has said that it is unlikely that the virus can be eliminated, and as a result, we must all learn to live with it. many workplaces remain closed for workers, and mental health issues caused by social isolation remain on the rise. Interior designers and architects have a responsibility to society and must draw from this emerging public health issue to invent creative solutions.

2 Methodology

This study employs descriptive research methods. The main methodology comprises a careful review of available literature and case applications to synthesize a resource for design considerations relevant to workplace design in the post COVID-19 era. In reviewing such design measures, consideration is dedicated to assessing whether the proposed modifications to workspace design norms are consistent with sustainable design concepts. Ultimately, this paper aims to describe the current state of affairs and recent advancements on workplace design requirements to combat the challenges of COVID-19 and inform future research on this topic.

3 Discussion

Companies that have allowed workers to work from home have found that it works well in many instances and that individuals are still productive. It is particularly beneficial for businesses that have suitable technology to enable work-from-home activities. People, on the other hand, need social contact, and meeting in person still has an emotional effect (Nediari et al. 2021). However, going to the workplace has benefits and advantages. Employees are able to interact with each another's body, facial, and voice language. This has the potential to foster trust and a sense of community among the participants, and can contribute to building a sustainable workplace. In particular, in-person time is essential for people who work in a team (Rafter 2020). Besides, there are still workers who lack the tools to work effectively from home, have difficulty accessing technology to perform their job, or reside in regions with no cellphone coverage. The companies are responsible for establishing the back-to-work policy, and employees must feel secure about it. Staff will come back to work one at a time, and important employees are returning first. An alternating work schedule that divides workers into shifts ensures that a constant number of employees are present in the workplace at the same time (Nediari et al. 2021). Despite the lack of journal references focused on how Covid-19 effects workplace interior design, the topic of making the path back to thc workplace has been discussed via a number of interviews, e-papers, and webinars from designers all over the globe. Health of the employees will be under the spotlight, and new measures to prevent infection and illness will be adopted. It is discovered that workplace design and furnishings have specific safeguards in place to protect workers and help keep the disease from spreading. There are many factors to consider while designing and creating a 'Covid-19 safe' office based on the information gathered (Nediari et al. 2021).

As a new vision of the workplace layout, a good solution could be a secure workplace with a safe workspace setting of Six Feet termed "Six Feet Office", where the minimum distance between employees should be two meters or six feet (Kretchmer 2021). It is capable of creating a new configuration for the workspace's layout furnishings. People will be seated back-to-back rather than face-to-face. The checkerboard or zigzag pattern will be used for the table layout. Due to space constraints, the size of the office desk has decreased, which has gone from 1.8 m to 1.6 m to the current 1.4 m. Employees may also have separate safety panels or barriers in between them. Physical barriers between workers may also help minimize workplace transmission if social distance isn't feasible. These barriers may be simple plastic sneeze shields or partitions between workstations (Mason 2021). The workstations arrangements shown in Figs. 1 and 2 would be more suited to dealing with the current concept of furniture layout after Covid-19. Before the pandemic, most offices had an open workspace plan that did not have any separating partitions or walls. It facilitates cooperation and communication across the company's departments. The open workplace encourages employees to connect more and bridges communication and distance gaps. This practice benefits the business, but since the pandemic, neighboring seat positions that

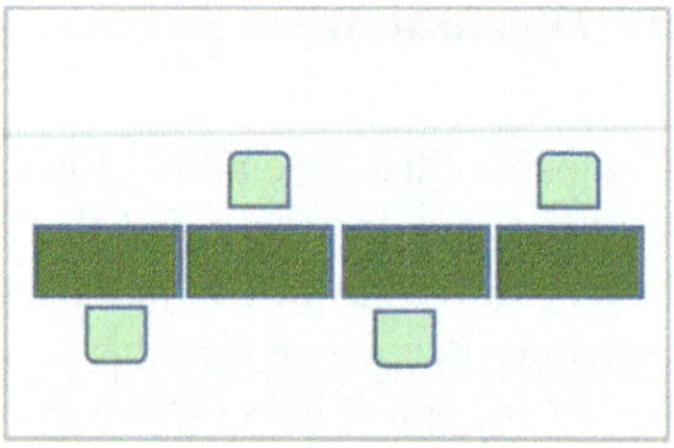

Fig. 1 Checkerboard pattern workstation layout (Nediari et al. 2021)

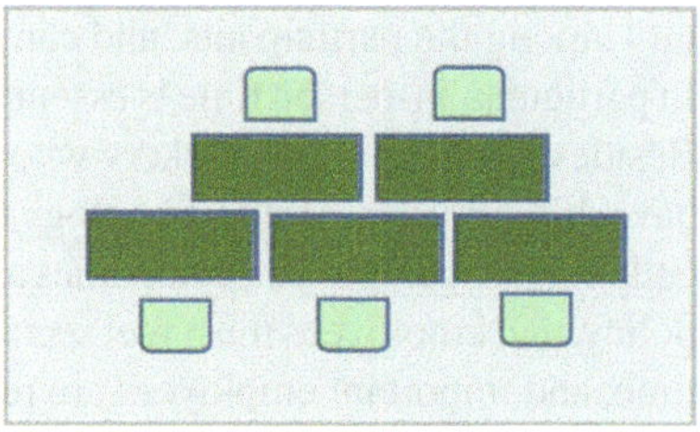

Fig. 2 Zigzag pattern workstation layout (Nediari et al. 2021)

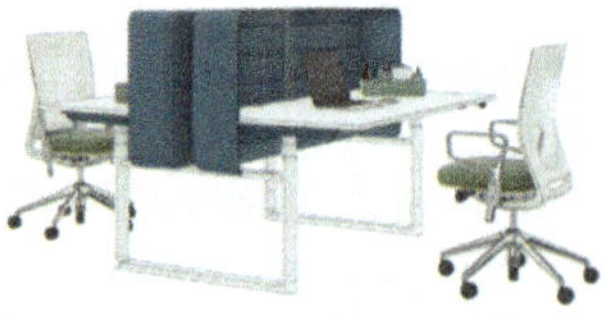

Fig. 3 Panels separating the workers in the workstation from Vitra (Nediari et al. 2021)

face each other must be changed to minimize the chance of Covid-19 transmission (Nediari et al. 2021).

Figure 3 shows a workstation designed by Vitra, a Switzerland office design company. It's one of the workstation designs that's been used after Covid-19. It's designed to be safe, and the high separation screen serves as a partition or a barrier between two employees. The Kinzo architectural design firm developed DisCo (Fig. 4), "transparent, mobile" partitions to minimize transmission between workers while encouraging social contact and cooperation (Kinzo 2020). The DisCo's color choices have been carefully chosen to prevent a sterile, clinical office environment and instead offer some vibrancy. However, the way light interacts with the colors of these dividers should be addressed (Mason 2021).

Other design ideas establish physical boundaries like the dreaded office cubicle. The 'Qworkntine Concept' by Mohamed Radwan is an office pod intended to minimize transmission (see Fig. 5) (Radwan 2021). The pod's ventilation and materiality have been intended to keep covid-19 out. However, these pods fully isolate the user, surpassing the old workplace cubicles. Office cubicles were criticized for limiting

Fig. 4 DisCo dividers from Kinzo (Mason 2021)

Fig. 5 'Qworkntine Concept' designed by Mohamed Radwan (Mason 2021)

natural light and social interaction by obscuring views of the rest of the space and windows. This harmed cognitive function and raised stress. Cubicles had such a negative effect that individuals disliked working in them even before research showed they were detrimental to wellness. The risk of these pods is that if they do not address the prior design faults of the cubicle, they may undermine the purpose of people returning to work (Mason 2021).

Making greater space in the workplace is another apparent way to reduce transmission. This relates to social distancing and the 2 m rule, and may imply greater space between desks and individual workstations, as well as larger doorways and hallways for safer walking. Furthermore, wider, less crowded surroundings enhance wellbeing and conceptual cognitive thinking (Wyatt 2017). Creating space may assist guard against the virus while also improving the atmosphere, boosting wellbeing and therefore productivity. Employers may need to lower capacity or locate additional separate or bigger locations to meet this shift. One potential criticism of this approach is that it may raise real estate expenses for businesses by requiring more square meters per person; however, with the assistance of evidence-based design, this may be seen as an investment in employees' health and wellbeing that would eventually result in improved productivity. However, others in the business think it would be advantageous to decentralize certain offices and relocate teams throughout a city or its suburbs. This may decrease the usage of public transportation, reducing transmission possibilities. Also, reducing a building's capacity reduces the chance of the virus spreading. This would allow businesses to shut down a smaller part of the firm, minimizing disruptions in the working and living of employees (Mason 2021). Furthermore, remote work on a regular basis may help to decrease office space density. Working from home will become more frequent. Flexible working hours has become available, ranging from shift work to 3-to-2 models. Three days at work and two days at home in a week is referred to as the 3-to-2 rule. As a result, enabling remote working will offer the companies access to a larger pool of talent from across the world (Nediari et al. 2021).

A good concept to use in the workspaces is one-way foot traffic, which means to move in one direction only. One-entrance and one-exit door may be used to accomplish this. It will avoid two-way traffic, which has a strong chance of being touched by accident or unintentionally. A corridor must be at least 1.2 to 1.5 m wide in order for two persons to walk across it. In order to ensure a suitable space between workers the internal infrastructure and the transit zones should be adapted (Nediari et al. 2021). In the lobby hall, as well as the standing place in the elevators, signage is used for guidance. As a visual instructional guide, the office's floor and walls of the workspace should be covered in signage. Signs may be used to remind workers to walk clockwise with the one-way circulation to reduce transmission, imitating what occurs in the hospitals during a pandemic (Nediari et al. 2021).

The companies also need to take an active part in improvement of hygiene standards at the workspace to avoid the spread of the virus in the office environment. It is about maintaining the health of workers and the productivity of the company (Nediari et al. 2021). In the future, workplaces will include more hygiene stations, which are low-cost treatments that may significantly reduce transmission. The 'Grand Hotel Sanitary Stations' by Saguez & Partners are stylized and opulent (see Fig. 6) (Saguez and Partners 2020). These hygiene stations are in public areas to urge people to wash their hands and eliminate the need to touch doors first (Mason 2021).

To promote free movement, companies must install self-opening doors and elevators that react to voice commands rather than touch buttons. To minimize disease transmission, they also may need to increase the investment in the new contactless technology system. Technology may be utilized to remind workers of the need of maintaining social distance (Nediari et al. 2021). Touchpoints must be redesigned or eliminated to maximize the effect of hygiene stations (Mason 2021).

More cleaning standards and procedures may be implemented, and the companies may do more to educate workers about how to clean the areas. Tables and chairs made of easy to clean and wipe materials such as Medium-Density Fiberboard (MDF), melamine, metal such as aluminum or steel, and High-Pressure Laminate (HPL) may be used (Nediari et al. 2021). Incorporating natural antibacterial materials like brass or copper, or other non-porous, wipe-clean materials, may mean

Fig. 6 Photograph of Saguez & Partners 'Grand Hotel Sanitary Station' (Mason 2021)

automating doors and taps. However, utilizing 'wipe clean' materials may produce an unnatural, antiseptic, hospital-like atmosphere. Artificial and unnatural surroundings may be harmful to people's health, according to evidence-based ideas like biophilic design and evolutionary psychology. In response, the necessity for virus prevention now exceeds other health concerns. But long-term and sustainable solutions need a more comprehensive approach (Mason 2021).

On the other hand, the key to preventing the spread of Covid-19 is proper ventilation. Although it is possible to do so by just opening a window, many workspaces are sealed. It needs discussion and collaboration with building management, as it has a significant effect on the design of the building and the regulations that apply to the occupants of the building (Nediari et al. 2021). Since the emergence of this virus, workplace ventilation has been a hot issue. Good ventilation lowers virus concentration in the air and therefore reduces the danger of airborne transmission, according to UK government health and safety standards (HSE 2020). This ventilation may be through open windows or recirculated air from an air conditioner. Unfortunately, office buildings, especially high-rise buildings, do not have operable windows, which means that mechanical ventilation is required to keep these environments comfortable. However, this air can be cleaned before it is recycled using "filtration systems, high-efficiency filters, and ultraviolet-based devices" (HSE 2020). Company relocation from high-rise buildings to structures with greater natural ventilation and fresh air may be encouraged. Creating more outdoor workplace areas in warmer regions has been suggested to decrease airborne transmission while increasing employee welfare, according to Kellert in 'Biophilic design in Practice' (Mason 2021).

Moreover, in certain workplaces, the number of seats in the lounge and conference room must be reduced. On the other hand, some other offices may still choose to interact with each other through video conference rather than meeting in person. People will be more concerned about sharing a workplace with others, thus there will be fewer unassigned seats (Wells 2020). The open office layout has grown popular as a design trend as well as a cost-cutting strategy known as sustainable design. The main idea of this design is to focus on natural light and the space between the desks (Gibbens 2021). In 2018, some researches were conducted on the open office idea, which was used by a number of companies. It was found that face-to-face contact has decreased by 70% while internet communication has risen. Another research found that in a crowded environment, the worry of infection made psychology stressful. For companies and designers, the issue of creating the post-Covid-19 safe workspace has become a major topic. The companies consider how they could improve their working environment. According to some experts, open space designs may be redesigned to provide more personal space and regular cleaning routines. Figure 7 shows a Vitra-inspired workplace layout design (Nediari et al. 2021). The e-paper is based on a project they did for their company. The image illustrates how the workplace layout before the Covid-19 pandemic had 64 employees and 371 square meters of office space with the furniture arrangement. There are 25 chairs at the workspace, all of which are in a face-to-face posture with no separation or partitions between the employees. Six people may sit next to each other in the meeting room as well as six people can sit side by side in the cafeteria. Vitra has suggested a new workspace layout based on the same area, as shown in Fig. 8, using post-Covid-19 safe design consider-

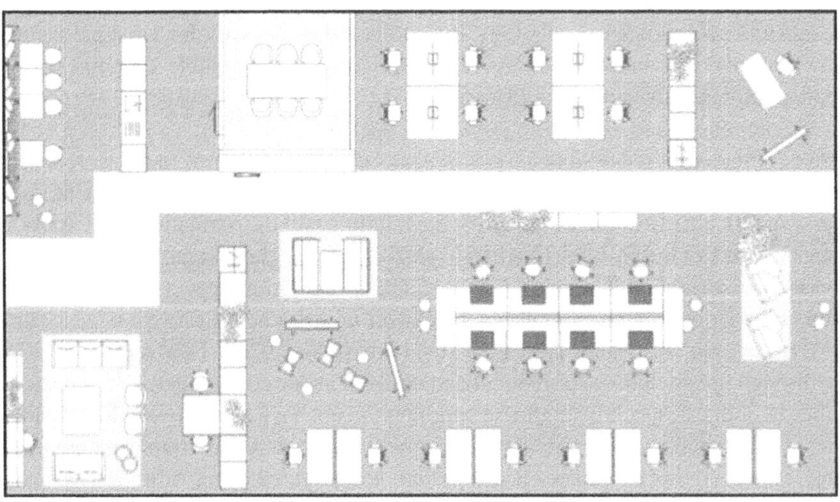

Fig. 7 Before the Covid-19 Pandemic, the existing furniture layout (Nediari et al. 2021)

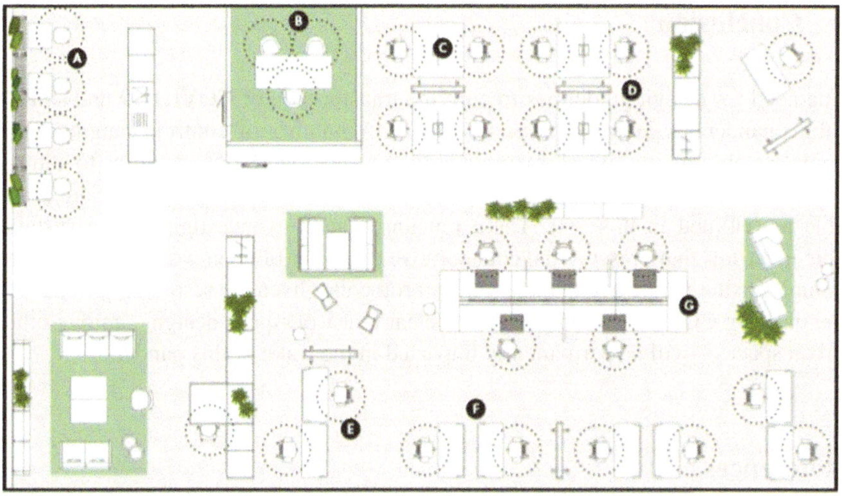

Fig. 8 Post-Covid-19 design proposed furniture layout (Nediari et al. 2021)

ations. The office plan, as shown in Fig. 7, is based on an open office workplace idea, with only a meeting room with a complete partition in the area (Nediari et al. 2021).

This is the post-Covid-19 safe office workspace layout modification. Various activities, including the cafeteria, meeting room, workplace, and management area, all have their own spaces. There is a dancing wall, which is a movable partition that could be used to split workspaces into zones in a variety of ways. It's made out of a metal frame with detachable whiteboards and pinboards that may be used as a shelf for books, TV unit, planter pot, or room partition (Hürlemann 2018). The workstation's capacity is 64 seats in Fig. 4, but it is decreased to 25 seats in Fig. 8 by adopting a zigzag and checkerboard layout. Additionally, as table dividers, the workstations have high separating screens. It is used in the six feet office to keep the social distance the workplace. To reduce disease transmission, the meeting area only has three chairs. There are still eight chairs in the cafeteria, and each table is separated from the others (Nediari et al. 2021). As long as the arrangement follows the 6 ft guidelines, the open-plan office layout is still appropriate for the post-Covid-19 design.

This issue is concerned with the sustainable design because the approach is focused on reducing the negative effect on the users and their environment. Increased standards of cleanliness and hygiene have a direct effect on the development of a healthy office workplace environment. The changes made to establish new workspace design guidelines supporting the post-Covid-19 safe office workplace (Nediari et al. 2021).

4 Conclusion

The need for design guidelines to curb the transmission of COVID-19 and similar future pandemics in workspaces is evident. A number of controls ranging from complete elimination of risk by emphasizing remote work opportunities to administrative controls that impact the flow of individuals in a given workspace are available as outlined in this text. There remains a need to investigate more sustainable measures that preserve the collaborative character of open workspaces without compromising public health and hygiene protocols. Given the resounding impact of the ongoing COVID-19 pandemic, it is clear that workspace design—among other urban spaces—will be permanently impacted in the wake of this pandemic.

References

Gibbens S (2021) Goodbye to open office spaces? How experts are rethinking the workplace. Science. https://www.nationalgeographic.com/science/2020/04/will-coronavirus-end-the-open-office-floor-plan/. Accessed 21 Sept 2021

Hse.gov.uk (2020) Ventilation and air conditioning during the coronavirus (covid-19) pandemic. https://www.hse.gov.uk/coronavirus/equipment-and-machinery/air-conditioning-and-ventilation/index.htm. Accessed 21 Sept 2021

Hürlemann S (2018) Dancing wall: official Vitra online shop. Vitra. https://www.vitra.com/en-us/product/dancing-wall. Accessed 21 Sept 2021

Kinzo-berlin.de (2020) Disco. Kinzo. https://kinzo-berlin.de/en/work/disco/. Accessed 21 Sept 2021

Kretchmer H (2021) Covid-19: is this what the office of the future will look like? World Econ Forum. https://www.weforum.org/agenda/2020/04/covid19-coronavirus-change-office-work-homeworking-remote-design/. Accessed 21 Sept 2021

Mason L (2021) Possible Futures of the Workplace: How has the Covid-19 pandemic amplified the need for a change towards an evidence-based, human-centred approach to office design?

Nediari A, Roesli C, Simanjuntak PM (2021) Preparing post covid-19 pandemic office design as the new concept of sustainability design. IOP Conf Ser Earth Environ Sci 729(1):012095. https://doi.org/10.1088/1755-1315/729/1/012095

Radwan M (2021) Qwork pods: winner of DNA Paris & BRONZE A' design awards. https://mradwan.net/portfolio/qworkntine/. Accessed 21 Sept 2021

Rafter D (2020) Does COVID-19 mean the end of the corporate office? Probably not (but do expect changes). REJournals. https://rejournals.com/does-covid-19-mean-the-end-of-the-corporate-office-probably-not-but-do-expect-changes/. Accessed 21 Sept 2021

Saguez & Partners (25 August 2020) Le bureau, moins mais mieux? Retour sur la conférence. Saguez & Partners. https://saguez-and-partners.com/2020/07/15/le-bureau-moins-mais-mieux/. Accessed 21 Sept 2021

Wells J (27 April 2020) After coronavirus: the office of the future is the office of the past. CNBC. https://www.cnbc.com/2020/04/27/after-coronavirus-the-office-of-the-future-is-the-office-of-the-past.html. Accessed 21 Sept 2021

Wyatt S (10 January 2017) Cubicles don't WORK. how architectural design affects your BRAIN | Scott Wyatt | TEDxSeattle. YouTube. https://www.youtube.com/watch?v=lFkJCpD0_V0&list=LL&index=13. Accessed 21 Sept 2021

The manufacturer's authorised representative in the EU is Springer Nature Customer Service Centre GmbH, Europaplatz 3, 69115 Heidelberg, Germany. If you have any concerns regarding our products, please contact ProductSafety@springernature.com

Printed and bound by CPI Group (UK) Ltd, Croydon, CR0 4YY
28/09/2026
02224093-0012